INFINITY

D0062383

INFINITY

The Quest to Think the Unthinkable

Brian Clegg

CARROLL & GRAF PUBLISHERS
New York

Carroll & Graf Publishers
an imprint of Avalon Publishing Group, Inc.
161 William Street
New York
NY 10038 2607
www.carrollandgraf.com

First published in the UK as *A Brief History of Infinity* by Robinson,
an imprint of Constable & Robinson Ltd, 2003

First Carroll & Graf edition 2003

Reprinted 2004

ISBN 0-7867-1285-6

Printed and bound in the EU

Library of Congress Cataloging-in-Publication Data
is available on file.

To Gillian, Rebecca and Chelsea,
and also to Neil Sheldon at the Manchester Grammar School
for having patience over the matter of frogs and lily-pads.

CONTENTS

ACKNOWLEDGEMENTS

Many thanks to all those who provided information and assistance in putting this book together including the staff at Swindon Central Library and the British Library, Forrest Chan of LiPACE (Open University of Hong Kong), Professor David Fowler of the University of Warwick, Dr Jeremy Gray of the Open University, Professor Dale Jacquette of The Pennsylvania State University, Professor Eberhard Knobloch of the Institut für Philosophie in Berlin, Professor Shaughan Lavine of the University of Arizona, Professor Rudy Rucker of San Jose State University, Peggy Brusseau and James Yolkowski.

Thanks also to my agent, Peter Cox, in his consistent efforts to encourage publishers that finance and the infinite are not incompatible, and to all those who have been involved at Constable & Robinson, particularly Jan Chamier and Pete Duncan.

1

TO INFINITY AND BEYOND

In this unbelievable universe in which we live there are no absolutes. Even parallel lines, reaching into infinity, meet somewhere yonder.

Pearl S. Buck, *A Bridge for Passing*

THE INFINITE IS A CONCEPT SO REMARKABLE, so strange, that contemplating it has driven at least two great mathematicians over the edge into insanity.

In the *Hitch-hiker's Guide to the Galaxy*, Douglas Adams described how the writers of his imaginary guidebook got carried away in devising its introduction:

'Space,' it says, 'is big. Really big. You just won't believe how vastly, hugely, mind-bogglingly big it is. I mean, you may think it's a long way down the street to the chemist, but that's just peanuts to space. Listen . . .' and so on. After a while the style settles down a bit and it starts telling you things you actually need to know . . . [1]

Infinity makes space seem small.

Yet this apparently unmanageable concept is also with us every day. My daughters were no older than six when they first began to count quicker and quicker, ending with a blur of words and a triumphant cry of 'infinity!' And though infinity may in truth make space seem small, when we try to think of something as vast as the universe, infinite is about the best label our minds can apply.

Anyone who has broken through the bounds of basic

mathematics will have found the little ∞ symbol creeping into their work (though we will discover that this drunken number eight that has fallen into the gutter is not the real infinity, but a ghostly impostor). Physicists, with a carelessness that would make any mathematician wince, are cavalier with the concept. When I was studying physics in my last years at school, a common saying was 'the toast rack is at infinity'. This referred to a nearby building, part of Manchester Catering College, built in the shape of a giant toast rack. (The resemblance is intentional, a rare example of humour in architecture. The companion building across the road, when seen from the air, looks like a fried egg.) We used the bricks on this imaginative structure to focus optical instruments. What we really meant by infinity was that the building was 'far enough away to pretend that it is infinitely distant'.

Infinity fascinates because it gives us the opportunity to think beyond our everyday concerns, beyond *everything* to something more – as a subject it is quite literally mind-stretching. As soon as infinity enters the stage, it seems as if common sense leaves. Here is a quantity that turns arithmetic on its head, making it seem entirely feasible that $1 = 0$. Here is a quantity that enables us to cram as many extra guests as we like into an already full hotel. Most bizarrely of all, it is quite easy to show that there must be something that is bigger than infinity – which surely should be the biggest thing there could possibly be.

Although there is no science more abstract than mathematics, when it comes to infinity, it has proved hard to keep spiritual considerations out of the equation. When human beings contemplate the infinite, it is almost impossible to avoid things theological, whether in an attempt to disprove or prove the existence of something more, something greater than the physical universe. Infinity has this strange ability to be many things at once. It is both practical and mysterious. Scientists and engineers use it quite happily because it works – but they consider it a black box, having the same relationship with it that most of us do with a computer or a mobile phone, something that does the job even though we don't quite understand how.

The position of mathematicians is rather different. For them, modern considerations of infinity shake up the comfortable, traditional world in the same way that physicists suffered after quantum mechanics shattered the neat, classical view of the way the world operated. Reluctant scientists have found themselves having to handle such concepts as particles travelling backwards in time, or being in two opposite states at the same time. As human beings, they don't understand why things should be like this, but as scientists they know that if they accept the picture it helps predict what actually happens. As the great twentieth-century physicist Richard Feynman said in a lecture to a non-technical audience:

> It is my task to convince you not to turn away because you don't understand it. You see, my physics students don't understand it either. That is because I don't understand it. Nobody does.[2]

Infinity provides a similar tantalizing mix of the normal and the counter-intuitive.

All of this makes infinity a fascinating, elusive topic. It can be like a deer, spotted in the depths of a thick wood. You will catch a glimpse of beauty that stops you in your tracks, but moments later you are not sure if you saw anything at all. Then, quite unexpectedly, the magnificent animal stalks out into full view for a few, fleeting seconds.

A real problem with infinity has always been getting through the dense undergrowth of symbols and jargon that mathematicians throw up. The jargon is there for a very good reason. It's not practical to handle the subject without some use of these near-magical incantations. But it is very possible to make them transparent enough that they don't get in the way. We may then open up clear views on this most remarkable of mathematical creatures – a concept that goes far beyond sheer numbers, forcing us to question our understanding of reality.

Welcome to the world of infinity.

2

COUNTING ON YOUR FINGERS

Alexander wept when he heard from Anaxarchus that there was an infinite number of worlds; and his friends asking him if any accident had befallen him, he returns this answer:

'Do you not think it a matter worthy of lamentation that when there is such a vast multitude of them, we have not yet conquered one?'

Plutarch, *On the Tranquillity of the Mind*

COUNTING A SEQUENCE OF NUMBERS, one after the other, is a practice that is ingrained in us from childhood. The simple, step-by-step progression of the numerals is so strong that it can be surprisingly difficult to break out of the sequence. Try counting from one to ten as quickly as you can out loud in French (or another language where you know the basics, but aren't particularly fluent). Now try to keep up the same speed counting back down from ten to one. The result is usually hesitation; the rhythm breaks as we fumble around for the next number. We are tripped up trying to untangle that deep-seated progression.

Number sequences have become embedded in our culture, often as the focus of childhood rhymes. The most basic of these are simple memory aids dating back to our first attempts to count:

One, two, buckle my shoe,
Three, four, knock on the door,

Five, six, pick up sticks,
Seven, eight, lay them straight,
Nine, ten, a big, fat hen,
Eleven, twelve, dig and delve,
Thirteen, fourteen, maids a'courting,
Fifteen, sixteen, maids in the kitchen,
Seventeen, eighteen, maids in waiting,
Nineteen, twenty, my plate's empty.

There's a degree of desperation in some of the later rhymes in the sequence, but also a fascinating reminder of a world that has disappeared in its imagery of shoe buckles and maids. The repetitious, hypnotic rhythm helps drive the number values into place.

Other doggerel is more oriented to singing than the basic chanted repetition of 'one, two, buckle my shoe', a typical example being

One, two, three, four, five,
Once I caught a fish alive,
Six, seven, eight, nine, ten,
Then I let it go again.

Equally valuable for practice with counting are songs like *Ten Green Bottles*, running downwards to make a child's grasp of the numbers more flexible.

But number rhymes aren't limited to helping us learn the basics of counting. More sophisticated verses add symbolism to sequence. It's difficult not to feel the power of numbers coming through in a rhyme like the magpie augury. This traditional verse form, familiar to UK children's TV viewers of the 1970s from its use in the theme song of the magazine programme *Magpie*, links the number of magpies (or occasionally crows) seen at one time with a prediction of the future. It's not so much about counting as about fortune-telling.

The TV show used the first part of a common sanitized version:

One for sorrow, two for joy,
Three for a girl and four for a boy,

Five for silver, six for gold,
Seven for a secret, never to be told,
Eight for a wish and nine for a kiss,
Ten for a marriage never to be old.[3]

But there's more earthy realism in this early Lancashire variant:

One for anger, two for mirth,
Three for a wedding and four for a birth,
Five for rich, six for poor,
Seven for a bitch [or witch], eight for a whore
Nine for a burying, ten for a dance,
Eleven for England, twelve for France.[4]

Many children develop a fascination with the basic sequence of counting numbers. Once youngsters have taken on board the rules for naming the numbers, it's not uncommon for their parents to have to beg them to stop as they spend an inordinate amount of time counting up and up. Perhaps their intention is to get to the end, to name the 'biggest number'. But this is a task where a sense of completeness is never going to be achieved. A child could count for the rest of his or her life and there would still be as many numbers to go. It seems that children are fascinated by the order, the simple pattern of such a basic, step-by-step number string.

It is part of human nature to like order, to see patterns even where no patterns exist. When we look at the stars, we imagine that they form constellations – shapes that link these points of light into a skeletal picture – where in reality there is no link between them. You only have to consider the Centaurus constellation in the southern sky. Its brightest star, Alpha Centauri, is the nearest to ours, a mere four light years distant; the next brightest in the constellation, Beta Centauri (or Agena), lies 190 light years away, more than 45 times more distant. We are mistakenly linking together two objects that are separated by around 1,797,552,000,000,000 kilometres.

Our own Sun is much closer to Alpha Centauri than Agena is, yet we would hardly think of the Sun and Alpha Centauri as forming a pattern. Alpha and Beta Centauri are no more connected than Houston and Cairo are simply because they lie near the same latitude. Our eyes and brains, looking for structure in the myriad of winking points in the sky, deceive us into finding patterns.

We look for patterns primarily to aid recognition. Our brains simplify the complex shapes of a predator or of another human face into patterns to enable us to cope with seeing them from different directions and distances. In the same way that we look for patterns in the physical objects around us, we appreciate patterns in numbers, and, of these, few are simpler and more easily grasped than the series of whole or counting numbers 1, 2, 3, 4, 5, 6, . . .

The ellipsis at the end of the sequence, that collection of three dots '. . .' is a shorthand that stretches beyond mathematics, though we have to be a little careful about how it is being used here. In normal usage it simply means 'and so on' for more of the same, but mathematicians, more fussy than the rest of us, take it specifically to mean 'and so on without *any* end'. There is no point at which you can say the sequence has stopped, it just goes on. And on. And on.

From the earliest days of pre-scientific exploration of both the natural world and the world of the mind, such chains of numbers were examined with fascination. They are prime inhabitants of the landscape of mathematics, as rich and diverse as any family of animals in the biological terrain. Some sequences are almost as simple as the basic counting numbers, for example doubling the previous number in the sequence to produce

$$1, \ 2, \ 4, \ 8, \ 16, \ 32, \ 64, \ 128, \ldots$$

But things don't have to be so neatly ordered. You can have sequences that vary in direction, building in a dance-like movement, alternating two steps forward and one step back:

$$1, \ 3, \ 2, \ 4, \ 3, \ 5, \ 4, \ 6, \ 5, \ 7, \ldots$$

Or each value can be the sum of the previous two, the so-called Fibonacci numbers:

$$0, \ 1, \ 1, \ 2, \ 3, \ 5, \ 8, \ 13, \ 21, \ldots$$

Moreover, we can move on from addition and subtraction to wilder excursions that multiply, taking off like a flock of birds surprised on a lake: the squares, the original counting numbers multiplied by themselves, for instance,

$$1, \ 4, \ 9, \ 16, \ 25, \ 36, \ 49, \ldots,$$

or the fast-accelerating progress of a sequence that multiplies the two previous entries,

$$1, \ 2, \ 2, \ 4, \ 8, \ 32, \ 256, \ 8192, \ldots$$

Most of these types of sequence were known by the Greek philosophers who first pondered the nature of numbers. But one particular class seemed particularly to have fascinated them. These were progressions, not of whole numbers, but of fractions.

The simplest fractional sequence takes each whole number and makes it the bottom part of the fraction:

$$1, \ \frac{1}{2}, \ \frac{1}{3}, \ \frac{1}{4}, \ \frac{1}{5}, \ \frac{1}{6}, \ldots$$

This sequence of numbers is not particularly special. If we were to add each term to the next, the total would grow without limit. But the Greek philosophers noticed a very different – a bizarrely different – behaviour when they made a tiny change. Instead of using the counting numbers in the bottom part of the fraction, the new sequence is formed by doubling the bottom part of the fraction. The result,

$$1, \ \frac{1}{2}, \ \frac{1}{4}, \ \frac{1}{8}, \ \frac{1}{16}, \ \frac{1}{32}, \ldots,$$

has a very strange property, strange enough for the philosopher Zeno to use it as the basis for two of his still-famous paradoxes.

We know very little directly of Zeno's work. All his writing was lost except for a few hundred words (and even the attribution of those to him is questionable). What remains is the second-hand commentary of the likes of Plato and Aristotle, who were anything but sympathetic to Zeno's ideas. We know that Zeno was a student of Parmenides, who was born around 539 BC. Parmenides joined the settlement at Elea in southern Italy. The ruins of this Phoenician colony can still be seen outside the modern-day Italian town of Castellammare di Velia. It was here that the Eleatic school pursued a philosophy of permanent, unchanging oneness – believing that everything in the universe is as it is, and all change and motion is but illusion.

For all we know, Zeno may have contributed much to Eleatic philosophy, but now he tends to be remembered as a mathematical one-hit wonder. What has come to us, even though only as a dim reflection in the comments of others, is his fascination with tearing apart the way we think of motion. This demonstrates a fundamental belief of the Eleatics, the denial of the existence of change, but even in this indirect form it seems possible to detect a metaphorical glint in Zeno's eye as he puts up his arguments. The later writers who pass on the paradoxes point out that this was a youthful effort, and though the intention of describing them this way was to be scornful, in fact there is a very positive element of youthful challenge in these ideas.

In all, forty of Zeno's reflections on the static universe have been recorded, but it is four of them that continue to capture the imagination, and it is these four that particularly impinge on the consideration of motion and of that strange sequence of numbers $1, \frac{1}{2}, \frac{1}{4}, \frac{1}{8}, \frac{1}{16}, \frac{1}{32}, \ldots$

The most straightforward of the four tells the story of Achilles and the tortoise. Achilles, arguably the fastest man of his day, the equivalent of a modern sports star, takes on the ponderously slow tortoise in a race. Considering the result of a rather similar race in one of Aesop's fables (roughly contemporary with Zeno's paradoxes), it's not too surprising that the tortoise wins. But unlike the outcome of the race between the tortoise and the hare, this unlikely result is not brought on by laziness and presumption. Instead it is the sheer mechanics of motion that Zeno uses to give the tortoise the winner's laurels.

Zeno assumes that Achilles is kind enough to give the tortoise an initial advantage – after all this is hardly a race of equals. He allows the tortoise to begin some considerable distance in front of him. In a frighteningly small time (Achilles is quite a runner), our athletic hero has reached the point that the tortoise started from. By now, though, however slow the tortoise walks, it has moved on a little way. It still has a lead. In an even smaller amount of time, Achilles reaches the tortoise's new position – yet that extra time has given the tortoise the opportunity to move on. And so the endless race carries on with the physical equivalent of those three dots, Achilles eternally chasing the tortoise but never quite catching it.

Another of the paradoxes, called the dichotomy, is very closely related. This shows that it should be impossible to cross a room and get out of it. Before you cross the room, you have to reach the half-way point. But you can't get there until you've reached one-quarter of the way across the room. And that can't be reached before you get to the one-eighth point – and so on. The sequence

$$\frac{1}{2}, \frac{1}{4}, \frac{1}{8}, \frac{1}{16}, \frac{1}{32}, \cdots$$

gives an insight into what is happening. You are never even going to reach the $\frac{1}{2}$, because you haven't first reached the $\frac{1}{4}$, because you didn't get to the $\frac{1}{8}$th marker, and so on – and you can never get started because you can't define the initial point. You can run down the sequence, splitting the distance further and further, forever. Which is the first point you reach? We can't say, and so, Zeno argues, you can never achieve motion.

It's easy to dismiss these two paradoxes by pointing out that

Achilles' steps do not get smaller and smaller, so, as soon as he is within a pace of the tortoise, his next stride will take him past it. The same goes for our escape from the room, but in reverse – the first step you take will encompass all of the smaller parts of the sequence up to whatever fraction of the way across the room you get to. But that misses the point of the story. Zeno, after all, was trying to show that the whole idea of motion as a continuous process that could be divided up as much as you like was untenable.

It is helpful to put Zeno's paradox of Achilles and the tortoise alongside our sequence

$$1, \frac{1}{2}, \frac{1}{4}, \frac{1}{8}, \frac{1}{16}, \frac{1}{32}, \ldots$$

When we consider Achilles and the tortoise, let's make the rather rash assumption that the tortoise moves half as fast as the athlete (maybe Achilles was having a bad day, or the tortoise was on steroids). Then in the time Achilles moved a metre, the tortoise would have moved half a metre. In the time Achilles made up that half metre, the tortoise would do an extra quarter. As Achilles caught up the quarter, the tortoise would struggle on an eighth. All of a sudden, those numbers are looking very familiar. But the parallel isn't all that makes it interesting. Because once we start adding up the numbers in that series, something strange happens. Let's look at the same series of fractions, but add each new figure to the previous one to make a running total. Now we have

$$1, 1\frac{1}{2}, 1\frac{3}{4}, 1\frac{7}{8}, 1\frac{15}{16}, 1\frac{31}{32}, \ldots$$

However far you take the series you end up with 1 and something more, something more that is getting closer and closer to another 1 (making 2 in all), but never quite getting there. It doesn't matter how far you go, the result remains less than 2. Think of the biggest multiple of 2 you can possibly imagine – we'll give it a made-up name, a 'thrupple'. Then by the time you reach it you will have

$$1 + \frac{\text{thrupple} - 1}{\text{thrupple}},$$

but not quite 2.

Giving a huge number an arbitrary name like 'thrupple' isn't quite as bizarre as it seems. The biggest number around with a non-compound name, the googol, doesn't just sound like a childish name, it actually was devised by a child. According to the story of its origin, American mathematician Edward Kasner was working on a blackboard at home in 1938 and for some reason had written out the number below. 'That looks like a googol,' said his nine-year-old nephew Milton Sirrota. And the name stuck. That part of the story seems unlikely – it's more credible that Kasner was simply looking for a name for a number bigger than anyone can sensibly conceive and asked young Milton for suggestions. Either way, a googol is the entirely arbitrary number:

10,000,000,000,000,000,000,000,000,000,000,
000,000,000,000,000,000,000,000,000,000,
000,000,000,000,000,000,000,000,000,000,000

or 1 with 100 noughts after it.

Back at the sequence $1, \frac{1}{2}, \frac{1}{4}, \frac{1}{8}, \frac{1}{16}, \frac{1}{32}, \ldots$, what we find, then, is that however many times we add an item to the list, we will never quite reach a total of 2. As soon as Achilles' stride breaks this apparent barrier the tortoise's lead is done for, but the series itself can't break out of its closer and closer approach to 2.

A modern equivalent of the translation of this sequence into a physical reality, an Achilles and the tortoise for the twenty-first century, would be to imagine a series of mirrors reflecting a particle of light, a photon, each mirror half the distance from the next, set in a spiral. The obvious paradox here is that however many mirrors you cram in, however many reflections you allow for, the light will only travel a limited distance. But there is another, more

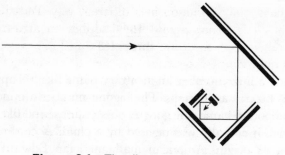

Figure 2.1 The disappearing photon.

subtle consideration. What happens to the photon at the end of the process? Where does it go? We know that after the first mirror it begins to spiral inwards, but light is incapable of stopping. The photon must continue travelling at 300,000 kilometres per second. So where does this particle go to?

In practice this seems to be one of those questions that have no meaning, because we would have to stray beyond the bounds of physical reality to reach the end result. Even if it were possible to divide up *space* into infinitely small chunks (we will return to the practicalities of this in the penultimate chapter), we know that physical matter isn't a continuous substance that can be divided forever, always coming up with a smaller version of the same thing. Reflection of light depends on an interaction between the photon of light and an electron in the material that it is being reflected off. Eventually, as the mirrors got smaller and smaller to fit into the remaining space they would have to become smaller than an atom, smaller than an electron – at which point reflection could no longer occur and the photon would continue on its journey without being further reflected.

In a moment we will return to that elusive sequence $1, \frac{1}{2}, \frac{1}{4},$ $\frac{1}{8}, \frac{1}{16}, \frac{1}{32}, \ldots$, which seems as if it should add up to 2 but never quite makes it, but first, for neatness, let's finish off Zeno and his paradoxes.

The other two mental pictures Zeno painted attack the

conventional view of motion in a different way. Perhaps most famous of all is Zeno's arrow. He describes an arrow, flying through space. After a certain amount of time has passed, it will have moved to a new position. But now let's imagine it at a particular instant in time. The arrow must be somewhere. You can imagine it hanging in space like a single frame from a movie. That's where the arrow is at, say, exactly ten minutes past two.

This is where the visual imagery of film comes in handy. There is now a video technique available that seems to make time stop. An object freezes in space as the camera pans around it, showing it from different directions. (In fact what is happening is that a series of cameras at different angles capture the moment, and their images are linked together by a computer to produce the illusion that the camera is panning.) Imagine that we do this for real. We stop time at that one instant and view the arrow.

Now let's do the same for another arrow that isn't moving at all. We won't worry too much about how this second arrow is suspended in space. If it's really a problem for you, we could work the paradox with two trucks, one moving and one stationary, but Zeno used an arrow, so I'd like to stick with that. The question Zeno asks is: how do we tell the difference? How does the *arrow* tell the difference? How does the first arrow know that it must change positions in the next moment, while the other, seemingly identical in our snapshot, stays still?

It's a wonderful problem – where Achilles and the tortoise can seem almost a matter of semantics, this one is a real poser. In fact, arguably it wasn't possible to truly answer Zeno until another great thinker also imagined bringing something to a stop. We have to leap forward 2,400 years to find Albert Einstein lying on a grassy bank, letting the sunlight filter through his eyelashes.

Relaxing on a summer's day at a park in the Swiss city of Berne in the early 1900s, Einstein imagined freezing a beam of light, not by taking a snapshot in time but by riding alongside it at the same speed. Now, as far as Einstein was concerned, the light was stopped. This is just the same as if you were in a car, and a truck was alongside you going at exactly the same speed – from your viewpoint the truck would not be moving. But Einstein's daydream of stopping light was a real problem, because the mechanism that the Scottish physicist James Clerk Maxwell had used to explain the workings of light some fifty years before would not allow it.

Maxwell's explanation of light depended on electricity and magnetism supporting each other in a continuous dance, an interplay that could only work at one speed, the speed of light. If it were possible for light to slow down, the delicate interaction of electricity and magnetism would collapse and the light would cease to exist.

If Maxwell was right, and Einstein assumed that he was, light could only continue to exist if it travelled at that one speed. And so Einstein made the remarkable leap of thinking that light would always move at that one particular speed, however fast you moved towards it or alongside it. Where we normally add speeds together when we move towards another moving object, or take speeds away from each other when we travel in the same direction as something else, light is a special case that won't play the game. This idea is at the heart of special relativity, which then, conveniently, makes Zeno's arrow less of a problem. Because it soon became apparent to Einstein that fixing light to a single speed (around 300,000 kilometres a second) changes the apparent nature of reality.

Einstein combined this fixed speed of light with the basic equations of motion that had stood unchanged since Newton's time. He was able to show that being in motion would change the appearance of an object. Everyday, apparently fixed, properties such as size, mass and the passage of time itself would seem different for an observer and the object that was observed. This effect is not very obvious until you get close to the speed of light, but it is always there. Einstein was able to show that moving at *any* speed with respect to something else changed both how you looked to that observer and how the observer looked to you.

Being in motion changes your world. And this provides a mechanism for the arrow to 'know' that it is in motion, because the world looks different in comparison with the world seen by a static arrow.

Relativity also deals with Zeno's final and least obvious paradox. This, called the stadium, imagines two rows of people passing each other in opposite directions. An athlete in one of these two rows will have passed twice as many body widths in the other row as he would have done if he had been running beside stationary people. Even though there is a physical limit to his speed, he seems to have run at twice that rate. In a way, what happens here is that our better understanding of relativity vindicates Zeno's views. He was using the example to show that the idea of moving at a certain speed is meaningless – and he was right. It is only ever possible to say we are moving at a certain speed relative to something else.

So with Zeno's four paradoxes in place, let's return to that sequence: $1, \frac{1}{2}, \frac{1}{4}, \frac{1}{8}, \frac{1}{16}, \frac{1}{32}, \ldots$ Is there a number of fractions we could add together that would bring the total up to 2? It would hardly be surprising if you said an infinite number. But what does that mean? If you ask children what infinity is when they are first introduced to it, they often say it is 'the biggest number that there can be'. Yet we've already said that, however many times you add on one of these decreasing fractions, you will never quite make 2. If infinity is 'the biggest number that there is', then surely we end up with a fraction that is

$$1 + \frac{\text{the biggest number} - 1}{\text{the biggest number}},$$

still not the round figure of 2.

This was a problem that proved an irritation for the ancient Greek philosophers who spent time thinking about such series. But to get into the frame of mind of a Zeno or Plato, we need to look at the numbers as the Greeks did themselves. The very wording of the sentence at the end of the previous paragraph would have been meaningless to the Greeks, because they did not have the same concept of a fraction that we do. In fact, numbers in general were handled in a very different way when the ancient Greek civilization flourished.

3

A DIFFERENT MATHEMATICS

If the doors of perception were cleansed every thing would appear to man as it is, infinite.

William Blake, *The Marriage of Heaven and Hell* (*A Memorable Fancy*)

IN MODERN TIMES WE TAKE FOR GRANTED two very specific tools of mathematics. One is being able to treat numbers as part of a number line – an imaginary horizontal ruler with 0 in the middle, stretching off into the distance in either direction with a series of markers along the way representing 1, 2, 3, etc., as you head to the right and the negative numbers to the left. Between each of the whole numbers sit all the fractions you'd care to imagine (and quite a few you wouldn't). For children in school today, the number line is an explicit part of their early learning of arithmetic. When subtracting, for example, they will often draw a number line and jump back down it by the appropriate value. For older readers, the number line might have been implicit, but it was still present in our education.

The second tool that comes into play when we work on the numbers from the number line is to take an algebraic approach – using a series of symbols that imply changing and combining the numbers (or place markers representing an undefined number) in various ways to come to a result. If, for example, we want to work out a price with sales tax from a basic price, we could have a table with every possible basic price on it and look up the appropriate value. Alternatively we could say that

$$\text{Price} = \text{basic price} \times (1 + \text{tax rate}).$$

This is a simple shorthand, saying that, to produce the price, we replace 'basic price' with a value, 'tax rate' with another value and carry out the calculation. This is, in essence, the secret of algebra. To keep things short (and unfortunately also making them less readable), mathematicians have a passion for using single letters (if necessary straying into Greek and other alphabets) for these labels, so they would probably say something like

$$p = b \times (1 + t),$$

but the effect is the same. Similarly, instead of having to explicitly write down how much energy every moving object has as a result of its speed, physicists say that this kinetic energy is $\frac{1}{2}mv^2$, where m is the mass of the object and v is its velocity. It doesn't matter what the values of m and v are, put them into the equation and it will work. Look at any mathematical paper written in the last 300 years and it is likely to depend upon the algebraic manipulation of many symbols in order to reach a conclusion.

The Greeks, it seems, at least at the height of their philosophical work, took a totally different approach. Their mathematics was not algebraic, but geometric. It was the mathematics of imagery.

Studies of the brain have shown that it can operate in two different modes, often labelled by association with its two halves – left-brain and right-brain thinking. Left-brain thinking is systematic, analytical and numerical. It is the principal approach of science and business and mathematics. Right-brain thinking is holistic, focusing on colour and imagery. It underlies the thought processes that are used in the arts. Ever since the development of the scientific method, the left-brain approach has dominated mathematics. Yet all of us are capable of both left- and right-brain thinking, and many people can handle information better from pictures than from words and numbers. Our modern processes tend to push us into left-brain mode, but for the ancient Greeks things were quite different. Mathematically at least, left-brain thought seems to have been alien to them. To perform arithmetic they thought in pictures.

With diagrams it was possible to handle concepts like equal to, larger than, smaller than, half the size of, twice the size of, and so on, without ever resorting to a number. Where the Greeks did use numbers mathematically they would not produce a string of digits 1, 2, 3, 4, 5, . . . as we would have, but rather a set of descriptions of objects corresponding to the numbers. In his book *The Mathematics of Plato's Academy*, David Fowler of the University of Warwick says we could think of the corresponding Greek series as

duet, trio, quartet, quintet, . . . [5]

(a solo item is omitted as it would have been treated separately from the larger numbers).

Similarly, the sequence we would write as

$$\frac{1}{2}, \frac{1}{3}, \frac{1}{4}, \frac{1}{5}, \dots$$

would effectively be formulated in words as

the half part, the third part, the quarter part, the fifth part, . . .

David Fowler points out that we should not mentally translate these into fractions – what the Greeks were thinking of by 'the half part', for instance, was a shape that was smaller than a unit item by a factor of 2. We not only have to think of the Greek 'fractions' visibly, as (for example) a square which is half the area of a rectangle twice the size, but also to remember that the relationship was seen in terms of whole numbers. What was observed was that the larger shape was twice the size of the smaller, not that the smaller was half the size of the larger.

If quantities were written down at all (diagrams were used

to put across much of the mathematical information that was required) the result would be a clumsy word phrase, all the more confusing as the convention of the time was to write with no gaps between the words. Reviel Netz points out in his *The Shaping of Deduction in Greek Mathematics* that to write $A + B = C + D$ (something that would not be literally done, as there was no algebraic concept of replacing an unknown number with a letter), the English equivalent of the Greek formula would be

THEAANDTHEBTAKENTOGETHERAREEQUALTOTHECANDTHED [6]

This doesn't mean that the ancient Greeks didn't have symbols for numerals. It would have made accounting very difficult if they were restricted to words, and in fact there were two approaches taken to writing numbers down. The older form, technically referred to as Acrophonic but sometimes called Attic, was a very simple approach. The number 1 was just a vertical bar I, with further bars added to count upwards. Five letters were used to stand in for bigger values to avoid too many bars being written: 5 was pi (Π – the symbol used is sometimes confused with gamma (Γ), as it was often written with a shorter right downstroke), 10 was delta (Δ), 100 eta (H), 1,000 chi (X) and 10,000 mu (M).

The term Acrophonic comes from the way the particular symbols were chosen – as each one was the first letter of the name of the relevant number. M for 10,000, for instance, was short for myriad (or, more precisely, for $\mu\acute{\nu}\rho\iota\omega$ – where μ is what we would now call the lower case version of M, mu).

To combine symbols, they could either be strung together, Roman fashion, so 42 would be $\Delta\Delta\Delta\Delta$II, or they could have a superscript as a multiplier. This would mean that to get 52 you would multiply 5 by 10 and add 2: Π^{Δ}II.

However, it is, perhaps, misleading to say that this was the early Greek number system, because there is strong evidence that it was hardly ever used for anything but inscriptions. It hasn't been found in works of mathematics. It seems likely that this Acrophonic system was used in early accounting – you can imagine a scribe ticking off items on a tablet – but it may never have been a tool of the mathematicians.

Like the (now) more familiar Roman numerals, the Acrophonic approach is very clumsy when it comes to dealing with larger numbers. A more flexible system was adopted, and was certainly in widespread use by the time Greek culture was centred on Alexandria. This used the letters of the Greek alphabet to represent numbers ($\alpha = 1$, $\beta = 2$, $\gamma = 3$, and so on), but hadn't got the convenience we have of using placing to represent tens, hundreds, and so on. This meant that yet more letters were called in to symbolize bigger numbers up to 900 (π, for instance, was 80). After that, a stroke was placed at the start of the number to take it above a thousand, and the capital mu (M) was maintained from the Acrophonic symbol for 10,000. Larger numbers still were achieved by writing a number above the M, which took the system up to a myriad myriad – 100,000,000. All in all, quite a mess. Things didn't get any easier when dealing with fractions. As we have already seen, these were thought of as whole number parts. A fifth wasn't really a fraction, so much as 'a thing of which the whole is five times bigger'. This meant that to get a fraction like $\frac{3}{5}$ you would have to have three lots of the fifth part. These parts usually used the standard number with an accent. A third part, for example, would be shown something like γ'. However, just in case things were seeming too easy, β' did not mean a half part, but two third parts. Bizarrely, a half had several special symbols, one like a lightning zigzag ㄅ, but none related to the alphabet. The standard order in which parts were listed was

$$\beta', \; \text{ㄅ}, \; \gamma', \; \delta', \dots,$$

putting the odd 'two third parts' character at the front of the sequence.

To write out a number that combines an integer with one or more parts, the integer would be written first, then a combination of parts. So, $8\frac{3}{4}$ would be represented by $\eta\text{ㄅ}\delta'$, combining 8, $\frac{1}{2}$

and $\frac{1}{4}$. (If you are familiar with the Greek alphabet and are wondering why η is 8 and not 7, some letters that had been dropped from the alphabet were still used as numbers. There was a 'double gamma', digamma F, which was used as 6 – double the value of gamma γ at 3.)

Of course, producing these combinations of fractions is not in itself trivial. *We* can work out that $\frac{1}{2}$ plus $\frac{1}{4}$ makes $\frac{3}{4}$ by turning the half into two quarters and adding the top parts of the fractions $(\frac{1}{4} + \frac{2}{4} = \frac{3}{4})$ – but we have to constantly remind ourselves that the Greeks weren't dealing with fractions as we know them. Without the easy visibility of the parts of our fractions, the Greeks resorted to the standard approach of anyone faced with a complex calculation – using a mechanical aid.

These days we would turn to a spreadsheet or a calculator; then, the solution was to consult laboriously worked-out tables providing the relationship between parts. Each fraction would be broken down into a set of components to enable manipulation.

When the Greeks were looking at an infinite series like the one behind Zeno's paradoxes,

$$\frac{1}{2}, \ \frac{1}{4}, \ \frac{1}{8}, \ \frac{1}{16}, \ \frac{1}{32}, \dots,$$

it is quite probable that this would be written in words as the half, the quarter, the eighth part, and so on, but it seems equally likely that any observations on the 'sum' of the series would have been made visually, as positions marked on a line or rectangle, rather than conceiving the sum as we do today – and this approach is just as effective (in fact, perhaps more so) in demonstrating the tendency towards a limiting value.

It's easy to be dismissive of this visual approach as crude compared to the surgical precision of algebraic equations, but it can still be extremely enlightening. I'd like to take you to a classroom in Rusholme on the outskirts of Manchester in the early 1970s. A mathematics lesson has just finished in which the teacher had discussed a strange problem. It involved a frog, hopping towards

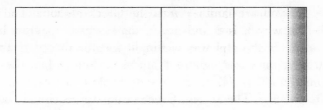

Figure 3.1 A visual representation of the series
$$1 + \tfrac{1}{2} + \tfrac{1}{4} + \tfrac{1}{8} + \tfrac{1}{16} + \tfrac{1}{32} + \cdots$$

a comfortable lily-pad for a rest. The frog was growing tired, so each hop was half the length of the previous one. Because of this, the frog would never, ever make it on to the lily-pad. The students weren't told it at the time, but it was a rewriting of one of Zeno's paradoxes, being used to illustrate one of the concepts behind the mathematical tool of calculus.

When the other students had left the room, one remained. He went up to the teacher and confessed that he just could not accept this idea. Surely if the frog hopped an infinite number of times, it would get to its destination, however small the hops became, wouldn't it? The mathematics teacher was a good teacher, and rather than dismiss this with 'it doesn't matter that you don't understand it, just use it', he gave some extra thoughts on the subject, and next day turned up with a book that went into much more detail on the relationship between the frog and lily-pad example and sequences of fractions. Yet despite all that help, despite the book, it wasn't until the student saw a visual representation, similar to Figure 3.1 above, that it all clicked into place.

I have to confess that the student was me. Taking a visual approach is not always a weakness. As we have seen, many of us think better visually than with numbers. It's why graphs can be much more powerful than tables of numerals at making financial data comprehensible. When the physicist Richard Feynman was working on the problem of explaining how light interacted with matter, his primary tool was not a complex equation but a (relatively) simple diagram, now called the Feynman diagram in his honour. This wasn't a way of simplifying the theory, quantum electrodynamics, for ignorant students, but rather the tool Feynman and his colleagues chose to explore and explain what was happening. The ancient Greek visual approach may not be ideal for much of mathematics, but it would be a severe misunderstanding to dismiss it as purely primitive.

To the ancient Greeks, dealing with these series of numbers was one thing, but infinity itself was quite a different matter. The nearest word they had to infinity was *apeiron*, but this was a term with quite different connotations to the feelings we now have for the infinite. It meant without bounds, out of control and untidy – *apeiron* was not a pleasant thing. Although it did not have the same meaning, *apeiron* had similar negative associations to those we now give to *chaos* in normal English (as opposed to the quite different, neutral mathematical definition).

It is easy to see why infinity caused the Greeks such discomfort – when some series are taken to the infinite extreme they produce very worrying results. Most of us can make the mental leap that will allow

$$1, \frac{1}{2}, \frac{1}{4}, \frac{1}{8}, \frac{1}{16}, \frac{1}{32}, \cdots$$

to gradually approach a sum of 2 when the components are added together. In effect, the sum would reach 2 if the number of items in the series ever actually made it to infinity. But what is the outcome of adding up the elements of the apparently even simpler sequence

$$1, -1, 1, -1, 1, -1, \ldots$$

At first glance, this will obviously add to 0, as each alternate entry in the sequence cancels out the next. But the result you get depends on how you do your pairing up. It's perfectly true that

$$(1 - 1) + (1 - 1) + (1 - 1) + \cdots$$

will add to 0 as each pair in brackets cancel each other out. But simply moving the brackets gets us

$$1 + (-1 + 1) + (-1 + 1) + (-1 + 1) + \cdots$$

Again, the brackets cancel themselves out, leaving not 0 but 1 as a total. The same series seems capable of adding up to 0 or 1.

This paradoxical result is sometimes transformed into the visual example of turning a light on and off an infinite number of times – does it end up off or on? Or is it in some strange indeterminate state, neither off nor on? Such a state of being is not as unlikely as it sounds. An exact parallel lies behind the curious condition of Schrödinger's cat, a hypothetical animal beloved of quantum physicists.

When the world is examined at the quantum level of individual particles of matter, normal expectations of behaviour go out of the window. Such particles, the electron for instance, can be in a number of 'either/or' conditions, rather like the way that a bit in a computer can be 0 or 1, or a coin can come up heads or tails when flipped. An example of such a condition is spin – an electron can be in one of two states, described as spin-up and spin-down (in other words, spinning clockwise or anticlockwise, depending on which way up you consider the axis round which it is spinning to be). Technically we don't know if the electron is actually spinning around, but it does have these two opposing states, and the label seemed one that was easy to visualize, so it stuck.

When a particle is produced, say by atomic decay, and examined, it will be found to be in one or other of these states randomly. But the strangeness of the quantum world means that, before it is examined, a particle, which could turn out to be in either state when measured, exists in both states simultaneously. This so-called superposition of states means that, for example, an electron would be spinning both ways simultaneously until it is observed, at which point the superposition is said to collapse and it becomes spin-up or spin-down.

Bemused by this concept, the Austrian physicist Erwin Schrödinger imagined a cat whose very existence depended on the state of a particle. His hypothetical experimental animal is put into a box with a lethal device that is triggered by a particle going into one of two available states. Until the box is opened, Schrödinger argued, if superposition truly exists and the particle is yet to be observed, then it is in both states at once – and that means that the cat should also be in both possible states.

This poor imaginary cat is simultaneously dead and alive until the box is opened and the superposed states collapse. Some argue that the device that checks the state of the particle is doing the observing, and so the superposed states are collapsed as soon as the lethal device is given the option of triggering. Others have suggested that this example demonstrates that the whole idea of superposed states is not a valid one. Yet the concept is part of the only consistently performing model that has yet been put forward to explain what is actually observed.

Similarly, it seems that, in the infinite series of switching on and off, the light bulb ends up in both states – and in this case there is no way to observe it and collapse the superposition, because we can never get to the end of the infinite series. (Realists will point out that the bulb will have blown long before an infinite number of switchings is reached, so it will definitely be off.)

It seems, then, that the sequence

$$1, \ -1, \ 1, \ -1, \ 1, \ -1, \dots$$

adds up to both 0 and 1, and unless we accept that $0 = 1$ it becomes obvious that infinity is more than a little tricky to deal with. The Greek answer was not to deal with it at all, or rather to hide it away in a fashion that made it unnecessary to think about it. Aristotle, whose views were also to shape so much of later Western scientific thinking, made a distinction on the matter of infinity that was to prove useful, but also was a fudge that made it possible to avoid the real issue for a couple of thousand years.

Unlike many of the Greek philosophers who would have a hand in the development of our idea of the infinite, we know a fair amount about Aristotle. He was born in 384 BC in Stagirus in northern Greece and at the age of 17 joined Plato's Academy in Athens, where he was to remain for 20 years.

The word 'academy' has become sufficiently part of the normal vocabulary that it is worth pointing out that this was the original. Plato, born in 427 BC in Athens, had founded his school when he left military service a couple of years before Aristotle's birth. Like many who have served in the military he was none too impressed with the capabilities of politicians, and set up his school with the specific intention of improving the quality of those in public life. The school, based in Athens, was situated in a grove of trees belonging to a man named Academos, hence the name 'Academy' (and the later, rather twee, expression 'the groves of academe'). By the time Aristotle arrived, the Academy was well established and respected. In fact it would remain operational until AD 529, a remarkable 900 years of existence – even the Universities of Oxford and Cambridge, the oldest English-speaking universities (schools were established at Oxford first, but both got their charters together in 1231), won't catch up until the 2130s.

Political upheaval seems to have made it wise for the then well-established Aristotle to move for a while to the island of Assos, before returning to Athens to set up his own school, the Lyceum. But it was while he was still at Plato's Academy that Aristotle wrote his *Physics* in which he was to try to tidy up the dangerous *apeiron*.

Aristotle introduced the idea of something being 'potentially infinite'. So when he counted up the sequence of whole numbers he accepted that it had no limit, because there was no point at which it stopped – but would not concede that it ever actually reached infinity, as this was an unreal concept. The world seemed to need something like infinity – time, after all, had no obvious end, and it seemed possible to take a straight line and divide it in half as many times as you like – yet by devising a potential infinity,

Aristotle could safely place the infinite outside of reality and simply head towards it.

This is not an arbitrary decision. Aristotle argues his reasoning very carefully. He begins by making it clear why the infinite is not just a matter of casual or even philosophical curiosity, but a practical piece of knowledge required for an understanding of nature:

> The science of nature is concerned with spatial magnitudes and motion and time, and each of these at least is necessarily infinite or finite . . . Hence it is incumbent on the person who specializes in physics to discuss the infinite and to inquire whether there is such a thing or not, and, if there is, what it is.[7]

Aristotle then begins his exploration of infinity by summing up the opinions of his predecessors and contemporaries. Both the Pythagoreans and Plato, he tells us, consider infinity to be an entity in its own right, rather than just an attribute of something else. The Pythagoreans, it seems, considered 'what is outside heaven' to be infinite, whereas Plato thought there was nothing outside the bounds of heaven and the infinite managed to fit within the universe. Others, those Aristotle refers to as the physicists, including Anaxagoras and Democritus (who came up with the forerunner of atomic theory), thought of infinity as an attribute of a substance, like its colour, rather than a true entity.

Why should people believe in infinity? Aristotle gives us five reasons. The simplest is based on his assertion that time is infinite, which he seems to take as incontrovertible. But, for Aristotle, one reason stands out above the rest:

> Most of all, a reason which is peculiarly appropriate and presents the difficulty that is felt by everybody – not only number but also mathematical magnitudes and what is outside the heaven are supposed to be infinite because they never give out in our thought.[8]

In other words, a strong point in favour of the existence of infinity is that we can't conceive of things like the series of counting numbers and the extent of the universe ever actually coming to a stop.

With these arguments in place, Aristotle begins to reveal his own views. The infinite can't be a thing, any more than number or magnitude can be a thing. Next he asks if a body can have this property – if there can be such a thing as an infinite body. This he counters by his definition of a body as something that is 'bounded by surface' – as infinity pretty well by definition cannot be completely bounded or it wouldn't be infinite, this doesn't make sense. (Aristotle couldn't say that infinity had no bounds at all, though. Take the series of whole, positive numbers. It is infinite, but it has a clear start, so it has at least one bound, its starting point.)

After a rather shaky discussion of why an infinite body, should there be such a thing, could neither be compound nor simple (and hence could not exist), Aristotle has proved pretty firmly from his viewpoint that you can't have a body that is infinite in size. But then he turns the argument back on itself. If there were no infinity, then time must have a beginning and an end, lines can't be indefinitely divided, numbers must stop. Aristotle presents us with a challenge:

> If, then, in view of the above considerations, neither alternative seems possible, an arbiter must be called in; and clearly there is a sense in which the infinite exists and another in which it does not.[9]

And now, at last, comes that subtle chimera, potential infinity, infinity that neither is nor isn't. Aristotle reminds us that the word 'is' could mean either something that is actually in existence, or something that has the potential to exist. Perhaps the answer to the problem of infinity is that it only potentially exists.

> But the phrase 'potential existence' is ambiguous. When we speak of the potential existence of a statue we mean that there will be an actual statue. It is not so with the infinite. There will not be an actual infinite. The word 'is' has many senses, and we say that the infinite 'is' in the sense in which we say 'it is day' or 'it is the games', because one thing after another is always coming into existence. For

> of these things too the distinction between potential and actual existence holds. We say that there are Olympic Games, both in the sense that they may occur and that they are actually occurring.[10]

It's a neat distinction, particularly in that last example. When we say 'there are Olympic Games', you could hardly argue with this statement. There *are* Olympic Games: they happen every four years with millions of people watching them. Yet unless we happen to be at the right point in the four year cycle, if a passing alien asked 'show me these Olympic Games of which you speak', we would have a problem. They do exist, but until the next Olympiad they are only potential. They are coming, but they aren't here. This was the nature of Aristotle's potential infinity – except the Infinite Olympiad never arrives. Aristotle ends Book 3 of the *Physics* with these words:

> This concludes my account of the way in which the infinite exists, and of the way in which it does not exist, and of what it is.[11]

The finality of this closing phrase, solidly putting a full stop on infinity and moving on to other things, would keep mathematicians happy all the way up to the nineteenth century. Whether infinity was real or unreal, comprehensible by the human or the divine, would occupy philosophers, but Aristotle's move of infinity into the virtual world of the potential made the mathematics work.

Even so, this dismissal of the ultimate did not stop others from working with larger and larger numbers, as if they were trying to find a way of describing the infinite itself.

4

THE POWER OF NUMBER

To see a World in a Grain of Sand,
And a Heaven in a Wild Flower,
Hold Infinity in the palm of your hand,
And Eternity in an hour.

William Blake, *Auguries of Innocence*, 1

WE HAVE SEEN HOW A CHILD gave an arbitrary (and let's face it, childish) name to the value

10,000,000,000,000,000,000,000,000,000,000,
000,000,000,000,000,000,000,000,000,000,
000,000,000,000,000,000,000,000,000,000.

A googol – one with a hundred zeros following. Although there are bigger named numbers derived from it – a googolplex is one with a googol noughts after it – there really is no need to give names to numbers of this scale. At one level, it's just a game, like the child's attempts to count higher and higher. But it's worth also bearing in mind that numbers themselves were considered to have significant power in ancient times, and to be able to name a large number could be thought of as a mark of authority.

For the Greek philosophers of the school of Pythagoras (he of the theorem that has dogged schoolchildren for hundreds of years), number became identified with creation. Specific numbers were given particular meanings. The school even had a numerical motto: *All is number*.

Pythagoras was born around 569 BC on the Ionian Greek island of Samos, set in the Aegean Sea. His father was a merchant, and the young Pythagoras travelled frequently with him. This made him more than usually receptive to the idea of travelling to Egypt when it was suggested in his early thirties that it would help his education – and it is probably there that Pythagoras was influenced in the direction that would eventually lead to the creation of his school at Croton in southern Italy.

The Pythagoreans considered numbers to be among the building blocks of the universe. In fact, one of the most central of the beliefs of Pythagoras' *mathematikoi*, his inner circle, was that reality was mathematical in nature. This made numbers valuable tools, and over time even the knowledge of a number's name came to be associated with power. If you could name something you had a degree of control over it, and to have power over the numbers was to have power over nature.

Particular properties were ascribed to each of the numbers from one to ten. These numbers had different personalities – some male, some female. Often the nature associated with each number was reflected in the shape of a pattern of dots made up from that number.

One, for example, stood for the mind – emphasizing its Oneness. Two was opinion, taking a step away from the singularity of mind. Three was wholeness (a whole needs a beginning, a middle and an end to be more than a one-dimensional point), and four represented the stable squareness of justice. Five was marriage – being the sum of three and two, the first odd (male) and even (female) numbers. (Three was the first odd number because the number one was considered by the Greeks to be so special that it could not form part of an ordinary grouping of numbers.)

This allocation of interpretations went on up to ten, which for the Pythagoreans was the number of perfection. Not only was it the sum of the first four numbers, but when a series of ten dots are arranged in the sequence 1, 2, 3, 4, each above the next, it forms a perfect triangle, the simplest of the two-dimensional shapes. So convinced were the Pythagoreans of the importance of ten that they assumed there had to be a tenth body in the heavens on top of the known ones, an anti-Earth, never seen as it was constantly behind the Sun. This power of the number ten may also have linked with ancient Jewish thought, where it appears in a number of guises – the ten commandments, and the ten *sefirot*, the components of the Jewish mystical cabbala tradition.

Such numerology – ascribing a natural or supernatural significance to numbers – can also be seen in Christian works, and continues in some new-age traditions. In the *Opus majus*, written in 1266, the English scientist-friar Roger Bacon wrote:

> Moreover, although a manifold perfection of number is found according to which ten is said to be perfect, and seven, and six, yet most of all does three claim itself perfection.[12]

Ten, we have already seen, was allocated to perfection. Seven was the number of planets according to the ancient Greeks, while the Pythagoreans had designated six as the number of the universe. Six also has mathematical significance, as Bacon points out, because if you break it down into the factors that can be multiplied together to make it – one, two and three – they also add up to six.

$$1 \times 2 \times 3 = 6 = 1 + 2 + 3.$$

Bacon reckoned that three was even more special because it was the only number that was the sum of all the parts of it that did divide into it (one) and the parts that aren't a factor (two). Arguably this is a bit of justification after the fact, with Bacon trying to show why three, representing the Holy Trinity of Father, Son and Holy Spirit, was so significant. Numerology not only commented on the significance of numbers but also used them as a means of divining the future, often converting words and names into number form and adding the results together to produce a prediction.

Yet the significance of numbers was not limited to these small, building-block values. According to Buddhist tradition, when Gautama Buddha was a young man, he had to name a whole series of numbers, starting in the millions and working up to figures comparable with a googol, with names for each step. Buddha was able effortlessly to name, for example,

$$100,000,000,000,000,000$$

as *achobya*, a name which in isolation has very little value (you are unlikely to go into a shop and ask for an achobya of anything), but which demonstrated his command over the impossibly large.[13]

Archimedes, writing at least 300 years earlier than the story of Buddha's numbering skills, also took the opportunity to plunge into the act of taming huge numbers, numbers that were of such a great size that they might well have been considered close to infinite. Despite the cumbersome Greek number system, which had no direct way of referring to any number greater than a myriad (10,000), Archimedes managed to estimate the number of grains of sand that would fill the universe.

We know little that is factual about Archimedes' life. A biography produced by an unknown writer Heracleides was mentioned in other Greek sources, but it no longer exists. (This could have been the same Heracleides that Archimedes refers to in one of his works, making him rather like an ancient Greek Boswell to Archimedes' Johnson). But Archimedes is thought to have died in the Roman attack on Syracuse, and if, as legend has it, he was around 75 at the time, this would put his birth date around 287 BC.

At the time of his death, Archimedes may well have been best known for the devices he planned and perhaps constructed. For him, according to the Roman writer Plutarch (writing a good 350 years later), these were 'diversions of geometry at play'[14] –

but the defenders of Syracuse might have seen things very differently. They were said to have used a range of mechanical engines devised by Archimedes to attack the Roman fleet, and even to have contemplated (but never built) a sun-powered death ray in the form of huge focusing mirrors to concentrate the Sun's light on distant Roman ships and set them alight.

But it was not Archimedes' undoubted mechanical genius, nor his detailed study of shapes and three-dimensional bodies that brings him into the story of infinity, but rather one of the strangest books in history. It is called the *Sand-reckoner*, and seems to be one of Archimedes' last works. It was written after he had returned to his native Syracuse from the time he is believed to have spent in Alexandria, and is addressed directly to the king of Syracuse, Gelon (or Gelo).

Archimedes seems to have had a close relationship with both Gelon and the previous king Hieron (there has been a suggestion that he was related to the royal family[15]), acting as a sort of interpreter of philosophical mysteries to the court. The *Sand-reckoner* opens with a challenge to the king to be more imaginative than many of his contemporaries:

> There are some, King Gelon, who think that the number of the sand is infinite in multitude; and I mean by the sand not only that which exists about Syracuse and the rest of Sicily but also that which is found in every region whether inhabited or uninhabited. Again there are some who, without regarding it as infinite, yet think that no number has been named which is great enough to exceed its magnitude. And it is clear that they who hold this view, if they imagined a mass made up of sand in other respects as large as the mass of the Earth, including in it all the seas and the hollows of the Earth filled up to a height equal to that of the highest mountains, would be many times further still from recognizing that any number could be expressed which exceeded the multitude of the sand so taken.
>
> But I will try to show you by means of geometrical proofs, which you will be able to follow, that, of the numbers named by me and given in the work which I sent to Zeuxippus, some exceed not only

the number of the mass of sand equal in magnitude to the Earth filled up in the way described, but also to that of a mass equal in magnitude to the universe.[16]

Fighting talk. And a serious challenge to the largely non-numerical Greek mathematics. Archimedes has first to define what he means by 'universe'. Our modern picture is a vast expanse of space, covering billions of light-years and filled with a huge collection of galaxies. At the time, the popular picture was of a (relatively) small sphere, centred on the Earth, containing nested spheres that held the planets in place with the stars in the final outer sphere. In passing, Archimedes tantalizingly refers to the first known suggestion that in fact the Sun rather than the Earth was at the centre of things, the 'heresy' for which Galileo would eventually be imprisoned. It is tantalizing because the work Archimedes refers to no longer exists, so it's rather as if the only view we had of a Shakespeare play was a description by a critic:

> Now you are aware that 'universe' is the name given by most astronomers to the sphere whose centre is the centre of the Earth and whose radius is equal to the straight line between the centre of the Sun and the centre of the Earth.
>
> This is the common account as you have heard from astronomers. But Aristarchus of Samos brought out a book consisting of some hypotheses, in which the premise leads to the result that the universe is many times greater than that now called. His hypotheses are that the fixed stars and the Sun remain unmoved, that the earth revolves around the Sun in the circumference of a circle, the Sun lying in the middle of the orbit, and that the sphere of the fixed stars, situated about the same centre as the Sun, is so great that the circle in which he supposes the Earth to revolve bears such a proportion to the distance of the fixed stars as the centre of the sphere bears to its surface.[17]

It is sometimes said that Archimedes dismissed the theory Aristarchus put forward, because his next line is:

> Now it is easy to see that this is impossible...[18]

But to be fair to Archimedes, all he is pointing out is that it isn't meaningful to say that the ratio of the size of the universe to the Earth's orbit is the same as the ratio of the centre of a sphere to its radius. After all, says Archimedes,

> since the centre of the sphere has no magnitude, we cannot conceive it to bear any ratio whatever to the surface of the sphere.[19]

However, rather than dismiss Aristarchus out of hand, instead Archimedes offers a practical interpretation of what Aristarchus had said:

> We must, however, take Aristarchus to mean this: since we conceive the Earth to be, as it were, the centre of the universe, the ratio which the Earth bears to what we describe as the 'universe' is the same as the ratio which the sphere containing the circle in which he supposes the Earth to revolve bears to the sphere of the fixed stars. For he adapts the proofs of his results to a hypothesis of this kind, and in particular he appears to suppose the magnitude of the sphere in which he represents the earth as moving to be equal to what we call the 'universe'.[20]

The language is more than a little convoluted, but reasonably logical. Archimedes supposes that Aristarchus meant that the size of his universe was in the same proportion to the size of the orbit of the Earth as the size of the traditional universe (the orbit of the Sun) was to the size of the traditional centre of the universe – the Earth itself. And it is this interpretation of Aristarchus' universe that Archimedes goes on to fill with virtual sand:

> I say then that, even if a sphere were made up of the sand, as great as Aristarchus supposes the sphere of the fixed stars to be, I shall prove that, of the numbers named in the *Principles*, some exceed in multitude the number of the sand which is equal in magnitude to the sphere referred to, provided that the following assumptions are made...[21]

Like all good mathematicians, Archimedes is about to tell us his 'givens', or axioms, the fundamental assumptions on which his theory is built. The book he refers to as the *Principles* is one that no longer exists, probably the one he mentions earlier as 'the work which I sent to Zeuxippus'. From the context it clearly dealt with the naming of numbers, but beyond that we know nothing about it.

To achieve an approximation of the size of the universe, Archimedes makes a total of four initial assumptions:

1. That the perimeter of the Earth is about 3,000,000 stades and not greater;

2. That the diameter of the Earth is greater than the diameter of the Moon but less than the diameter of the Sun;

3. That the diameter of the Sun is about 30 times greater than the Moon; and

4. That the diameter of the Sun is greater than the side of the chiliagon (a polygon with 1,000 equal sides) inscribed in the greatest circle in the sphere of the universe.

Let's consider the validity of those assumptions. Each stadium of measurement was around 180 metres in length (we still occasionally refer to dimensions in terms of stadium-like measurements, such as 'around the size of a football pitch'). So his guess that the perimeter of the Earth is 'about 3,000,000 stades' makes the Earth's circumference around 54,000 kilometres. This isn't bad at all, bearing in mind that the kilometre was originally defined as 1/10,000th of the distance from the pole to the equator through Paris – making 40,000 kilometres a good round figure for this.

Archimedes' second assumption, that the Earth was bigger than the Moon but smaller than the Sun was, of course spot on – and not totally obvious when you consider that the Sun and the Moon have almost identical apparent diameters in the sky (otherwise solar eclipses would not be anywhere near as spectacular as they are).

Things get a little more shaky on the third assumption that the diameter of the Sun is about 30 times that of the Moon. In fact, the Moon is around 3,480 kilometres in diameter (a quarter the size of the Earth), while the Sun measures up at 1,392,000 kilometres across, 400 times the size of the Moon. It is hardly surprising that Archimedes got it wrong – the scale is quite incredible – but even so, it only made him a factor of ten out.

The final assumption was this rather strange assertion that the diameter of the Sun is greater than the side of the chiliagon inscribed in the greatest circle in the sphere of the universe. In other words, it was assumed that the whole distance around the Sun's orbit (remember this was assuming a 'traditional' universe of the Sun's orbit of the Earth) was (roughly) no more than 1,000 times bigger than the diameter of the Sun. Again, this was surprisingly good, as long as you reverse the relationship between the Sun and the Earth.

Archimedes based this assumption on one report and one experiment. Aristarchus, he tells us, 'discovered that the Sun appeared to be about 1/720th part of the circle of the zodiac'.[22] Archimedes then improved on Aristarchus' measurement by taking a long rod with a small cylinder or disc fixed to the end. Just after sunrise, when he thought it was safe to look at the Sun (it never is, but at least it reduced the risk), he held up the rod and moved it backwards and forwards until the disc just covered the Sun's surface. He found that the angle the Sun's disc made was between 1/164th and 1/200th of a right angle, and from this, with reasonably simple geometry, he was able to come up with his estimate.

In reality, the Earth's orbit around the Sun (the closest we can come to Archimedes' 'sphere of the universe' averages around 939,000,000 kilometres, making the orbit around 675 times the size of the Sun, comfortably close to Archimedes factor of 1,000. Of course, this size bears no resemblance to the true magnitude of the solar system, let alone the true universe.

With those assumptions in place, Archimedes could show geometrically that his universe was no more than a myriad times the size of the Earth. He then somewhat arbitrarily decided that there are no more than a myriad grains of sand in the size of a poppy seed, and that a poppy seed (a conveniently small item) is not less than 1/40th of a finger's breadth in size. Now he can get on to counting those sand grains. But just writing it out in the Greek format would be an impractical task. Archimedes has to devise a new system of numbers that could cope with such a huge quantity.

With a myriad as the biggest named numbers, it was easy enough to describe a myriad of myriads – 100,000,000. Rather than give 100 million a specific name, he described all the numbers up to 100,000,000 as numbers of the first order. A myriad myriads then became the unit for a second order, so the biggest number in the second order was this unit of units, what we would represent as $100,000,000^2$, namely $100,000,000 \times 100,000,000$, or 10,000,000,000,000,000. The highest number in the second order became the unit of the next, going up to $100,000,000^3$, and so it went on, all the way to 100,000,000 multiplied by itself 100,000,000 times.

But that wasn't enough for Archimedes. The whole set of numbers up to this enormous value, he described as the first period, forming a next level of the hierarchy. The next period began with this number and went up to 100,000,000 times bigger. And so on. Just to write out the last number of the first period would require 1 followed by 800,000,000 zeros – this is no toy numbering system. Archimedes then worked out a relatively simple mechanism for performing arithmetical operations on these orders. Finally, he was ready to work upwards, counting the grains of sand in a sphere of a finger's breadth (no more than 10 units of the second order), and so on, up in leaps of a hundred, through a stadium to the diameter of the traditional universe.

He had already assumed that this universe, based on the imagined orbit of the Sun, was no bigger than 10,000,000,000 stades, which left the number of grains of sand to fill the universe as no more than 1,000 units of the seventh order – pretty trifling in the grand scheme of his numbering system. If you preferred Aristarchus' bigger universe, it was still a mere 10,000,000 units of the eighth order. To wrap it up, Archimedes comments:

> I conceive that these things, King Gelon, will appear incredible to
> the great majority of people who have not studied mathematics,
> but that to those who are conversant therewith and have given
> thought to the question of the distances and sizes of the Earth, the
> Sun and Moon and the whole universe, the proof will carry convic-
> tion. And it was for this reason that I thought the subject would be
> not inappropriate for your consideration.[23]

It would seem from Archimedes' words that Gelon was more
familiar with mathematics than many kings. But the conclusion
of the *Sand-reckoner* doesn't answer why Archimedes should want
to number the grains of sand that would fill the universe, or
devise a numbering system that went so far beyond the needs of
such a problem. Although we have no evidence to back this up, it
is quite probable that Archimedes was responding to the loose
remarks of those who would quite happily refer to the number of
grains of sand on a beach as infinite – to make it clear that these
were only trifling numbers compared with the numbers that
mankind (and Archimedes in particular) could name – putting
them nowhere near that uncomfortable concept, the infinite.

It also gave Archimedes a chance to play with astronomical
ideas in a way that hadn't really been done up to then. After all,
his father was himself an astronomer. When considering the
diameter of the Sun, Archimedes comments in the only known
reference to his father:

> Eudoxus declared [the diameter of the Sun] to be nine times as
> great [as the diameter of the Moon], and Pheidias my father twelve
> times.[24]

It is interesting that Archimedes then goes on to say that
Aristarchus thought this ratio to be between 18 and 20 times,
while

> I go even further than Aristarchus, in order that the truth of my
> proposition may be established beyond dispute, and I suppose the

diameter of the Sun to be about 30 times that of the Moon and not greater.[25]

Perhaps part of the reason for this whole book was not so much to deny the outlandish ideas of Aristarchus but to support them – to show that moving from a traditional universe to that of Aristarchus did not require a huge change in thought, merely involving a movement in size from the seventh order to the eighth. Whatever, Archimedes had mapped out some seriously large numbers.

Although the *Sand-reckoner* would remain something of a curiosity, references to it do crop up in poetic writing. John Donne, the great English poet of the seventeenth century, at least twice referred to Archimedes' efforts, making the apparently huge quantity Archimedes had named small in comparison to aspects of God, and in this particular Lenten sermon, the endlessness of suffering in hell:

> Men have calculated how many particular graines of sand, would fill up all the vast space between the Earth and the Firmament: and we find, that a few lines of ciphers will designe and express that number... But if every grain of sand were that number, and multiplied again by that number, yet all that, all that inexpressible, inconsiderable number, made up not one minute of this eternity...[26]

It's a very graphic picture of Donne's imagining of an eternity of suffering. In another sermon, this time from Easter Day, he compares the limitations of arithmetic as he knew it (no doubt devoid of infinity) to the limitlessness of God:

> How barren a thing is Arithmetique! (and yet Arithmetique will tell you, how many single graines of sand, will fill this hollow Vault to the Firmament) How empty a thing is Rhetorique! (and yet Rhetorique will make absent and remote things present to your understanding) How weak a thing is poetry! (and yet Poetry is a counterfait Creation, and makes things that are not, as though they were) How infirme, how impotent are all assistances, if they be put to express this Eternity.[27]

Donne might have limited sand grains to his sermons – but a successor was to bring them into poetry. Archimedes' attempts to deal with huge numbers seem reflected in the opening lines of one of William Blake's best-known poems. It was written in the first years of the nineteenth century, but not published until 1863, some 36 years after Blake's death, in a collection of his work by Dante Gabriel Rossetti.

The author of *Tyger, Tyger* and *Jerusalem* demonstrates his distaste for man's ability to abuse nature, and produced an augury that is significant for us not as a counting rhyme like the magpie auguries that appear on page 6, but rather in its use of a sand grain and infinity among the images that emphasize how the small actions of man impact all the vastness of creation. I've gone beyond the usual first four lines of *Auguries of Innocence* below to show the context of that first stanza that is so often plucked out in isolation:

To see a World in a Grain of Sand
And a Heaven in a Wild Flower,
Hold Infinity in the palm of your hand
And Eternity in an hour.

A robin redbreast in a cage
Puts all Heaven in a rage.
A dove house fill'd with doves and pigeons
Shudders Hell thro' all its regions.
A dog starv'd at his master's gate
Predicts the ruin of the state.
A horse misus'd upon the road
Calls to Heaven for human blood.
Each outcry of the hunted hare
A fibre from the brain does tear.
A skylark wounded in the wing,
A Cherubim does cease to sing.
The game cock clipp'd and arm'd for fight
Does the rising Sun affright.
Every wolf's and lion's howl
Raises from Hell a human soul.[28]

For Blake – and Donne before him – it was not only the attempt to count the grains of sand in the universe but the infinite itself that was linked to God. There was an almost inevitable juxtaposition of the ultimate in number and the ultimate reality, the creator of everything. This was no new concept, as we shall see in the next chapter.

5

THE ABSOLUTE

For the past and boundless eternity during which God abstained from creating man is so great, that, compare it with what vast and untold numbers of ages you please, so long as there is a definite conclusion of this term of time, it is not even as if you compared the minutest drop of water with the ocean that everywhere flows around the globe.

St Augustine, *The City of God*, Book XII.12

BLAKE'S RELIGIOUS BELIEFS were never entirely orthodox, but it would not be surprising if his concept of infinity embraced God or even if he had equated the infinite with God. It is a very natural thing to do. If you believe in a divine creator who is more than the universe, unbounded by the extent of time, it's hard not to make a connection between this figure and infinity itself.

There have been exceptions, philosophers and theologians who were unwilling to make this linkage. Such was the ancient Greek distaste for infinity that Plato, for example, could only conceive of an ultimate form, the *Good*, that was finite. Aristotle saw the practical need for infinity, but still felt the chaotic influence of *apeiron* was too strong, and so came up, as we have seen, with the concept of potential infinity – not a real thing, but a direction towards which real numbers could head. But such ideas largely died out with ancient Greek intellectual supremacy.

It is hard to attribute the break away from this tradition to one individual, but Plotinus was one of the first of the Greeks to make a specific one-to-one correspondence between God and the infinite. Born in AD 204, Plotinus was technically Roman, but was so strongly influenced by the Greek culture of Alexandria (he was born in the Egyptian town of Asyut) that intellectually, at least, he can be considered a Greek philosopher. He incorporated a mystical element (largely derived from Jewish tradition) into the teachings of Plato, sparking off the branch of philosophy since called Neoplatonism – as far as Plotinus was concerned, though, he was a simple interpreter of Plato with no intention of generating a new philosophy.

He argued that his rather loosely conceived god, the *One*, had to be infinite, as to confine it to any measurable number would in some way reduce its oneness, introducing a form of duality. This was presumably because once a finite limit was imposed on God there had to be 'something else' beyond the One, and that meant the collapse of unity.

The early Christian scholars followed in a similar tradition. Although they were aware that Greek philosophy was developed outside of the Christian framework, they were able to take the core of Greek thought, particularly the works of Aristotle and Plato, and fix it into a structure that made it compatible with the Christianity of the time.

St Augustine, one of the first to bring Plato's philosophy into line with the Christian message, was not limited by Plato's thinking on infinity. In fact he was to argue not only that God *was* infinite, but could deal with and contain infinity.

Augustine is one of the first Christian writers after the original authors of the New Testament whose work is still widely read. Born in AD 354 in the town of Tagaste (now Souk Ahras in Algeria), Augustine seemed originally to be set on a glittering career as a scholar and orator, first in Carthage, then in Rome and Milan. Although his mother was Christian, he himself dabbled with the dualist Manichean sect, but found its claims to be poorly supported intellectually, and was baptized a Christian in 387. He intended at this point to retire into a monastic state of quiet contemplation, but the Church hierarchy was not going to let a man of his talents go to waste. He was made a priest in 391 and became Bishop of Hippo (now Annaba or Bona, on the Mediterranean coast) in 395.

The Roman Empire that Augustine knew was the sad relic of an earlier age, already poised for collapse. In 410, Rome itself fell to the attack of Alaric the Visigoth, a barbarian who, it was thought, should not have stood a chance against the might of Rome that had stood unconquered for so long. Rather than accept the decline that had been under way for so many years, some commentators blamed the new official religion, Christianity. It was only when these weak Christians had taken over, it was argued, that Rome was lost.

Although Rome itself had fallen, the Roman hierarchy in the more remote areas was initially relatively unaffected. An imperial commissioner, Marcellinus, who was visiting the region, asked Augustine to write a response to the negative feelings that Christianity was generating – to make the people both more aware of the nature of Christianity and how its superiority over the displaced Roman religions made it anything but a suitable scapegoat for the fall of Rome.

The result was a long, sprawling series of books that came together over around 15 years as *The City of God*. By the time this collected work was complete, Augustine was already a respectable 71 (he died five years later). The book, published in parts as each was written, became an immediate success. It received what must be one of the first extant laudatory reviews, from Macedonius, Vicar of Africa. Macedonius praises the book for its

> happy combination of erudition and argument, doctrine and eloquence[29]

As Augustine built on his initial theme, *The City of God* became much more than a response to doubters, concerned by the fall of Rome. The final collection of 22 books is rather an attempt to pull together a total view of Christianity as it was then understood. To the viewpoint of a modern Christian it might seem odd that Augustine should concern himself at all with numbers, but it should be remembered that he was strongly influenced by Plato and took on board Plato's theories on the significance of certain numbers.

Augustine's numerical quest began not with infinity but with the number six. He pointed out that in the book of Genesis creation took six days to complete, yet this duration couldn't possibly have been a matter of necessity. This thought is quite logical, though not obvious when we are thinking in our normal, finite way of the time it takes us to complete a task. But, as Augustine emphasized, there is no particular reason why God would 'require a protracted time'. If we are to consider the point logically, an all-powerful God should be able to undertake the whole act of creation in an instant. Instead, Augustine argued:

> These works are recorded to have been completed in six days... because six is a perfect number... because the perfection of the works was signified by the number six. For the number six is the first to be made up of its own parts, i.e., of its sixth, third and half, which are respectively one, two and three, and which make a total of six.[30]

What Augustine meant here, as Roger Bacon later echoed, is that the only numbers that divide into six (its factors) are one, two and three. Multiply these factors together and you get six. Add them together and you also get six. Such numbers, made up of the sum of their factors had been termed a perfect number ever since the time of the Greeks. Unless you consider the number one, which is always something of an oddity, six is the smallest value exhibiting this 'perfect' quality.

Augustine, though happy to use such 'science' in the further-ance of his work, was quick to admit that this was not a specialist area, and so he had to use what he said with care. The modesty with which he points this out is touching. After talking about six, he also said that seven is a perfect number, but for a different reason. He went on:

> Much more might be said about the perfection of the number seven, but this book is already too long, and I fear lest I should seem to catch at an opportunity of airing my little smattering of science more childishly than profitably.[31]

Later on, in the next book of *The City of God*, Augustine began to work towards things eternal. He started by looking at different theories of the origin of the world that were current at the time, some of which have surprisingly strong echoes in present-day cosmological theories, from multiple parallel universes to a cyclically reborn universe:

> There are some, again, who though they do not suppose that this world is eternal, are of opinion either that this is not the only world, but that there are numberless worlds, or that indeed it is the only one, but that it dies, and is born again at fixed intervals, and this times without number.[32]

From this he moved on to put God alongside the concept of infinity. Some people, he argued, say that almost by definition, the infinite can't be handled by any mode of knowledge. Even God, they suggest, has only finite conceptions of his creations. Again, Augustine resorted to numbers to help out with this argument:

> As for their other assertion, that God's knowledge cannot comprehend things infinite, it only remains for them to affirm, in order that they may sound the depths of their impiety, that God does not know all numbers. For it is very certain that they are infinite; since, no matter at what number you suppose an end to be made, this number can be, I will not say increased by the addition of one more, but however great it be, and however vast be the multitude of which it is the rational and scientific expression, it can still be not only doubled, but even multiplied. Moreover, each number is so defined by its own properties that no two numbers are equal. They are therefore both unequal and different from each other; and while they are simply finite, collectively they are infinite.[33]

Augustine had asserted quite effectively that the numbers (by which he meant the counting numbers) go on forever – however big a number is, you can always find a bigger one, double it, or square it. Next he set God to work at the virtual abacus:

> Does God, therefore, not know numbers on account of this infinity; and does His knowledge extend only to a certain height in numbers, while of the rest He is ignorant? Who is so left to himself as to say so? Yet they can hardly pretend to put numbers out of the question, or maintain that they have nothing to do with the knowledge of God; for Plato, their great authority, represents God as framing the world on numerical principles; and in our books also it is said to God, 'Thou hast ordered all things in number, and measure, and weight.' ... Far be it, then, from us to doubt that all number is known to Him 'whose understanding,' according to the Psalmist, 'is infinite'.[34]

While Augustine's quote from Psalm 147 fits with a modern translation, it might be a little ingenuous as an argument, in that it depends on what the psalm writer had in mind by the word that is translated as 'infinity'. Similarly it would be easy to dismiss his other biblical reference as non-literal, a literary turn of phrase rather than literal description. Even so, there is considerable strength in Augustine's argument – if God *does* exist, it is hard indeed to envisage that such an unlimited creator could not conceive of anything bigger than a certain number (whatever that number happens to be).

Augustine then turned the numerical argument back on those who imagine that there is some sort of cyclical creation, where the universe is born and dies time and again throughout creation. This view, he said, was based on the inability to see how God could manage to work through an infinite, timeless period before the universe came into existence, somehow seeing ahead to the point when creation would be required:

> The infinity of number, though there be no numbering of infinite numbers, is yet not incomprehensible by Him whose understanding is infinite. And thus, if everything which is comprehended is defined

or made finite by the comprehension of him who knows it, then all infinity is in some ineffable way made finite to God, for it is comprehensible by His knowledge.[35]

Later heavyweight theologians would pull back a little from Augustine's certainty that God was able to deal with the infinite. While God himself was in some senses equated with infinity, it was doubted that he could really deal with infinite concepts other than Himself, not because he was incapable of managing such a thing, but because they could not exist. Those who restricted God's imagination in this way might argue that he similarly could not conceive of a square circle, not because of some divine limitation, but because there simply was no such thing to imagine. A good example is the argument put forward by St Thomas Aquinas.

Aquinas, born at Roccasecca in Italy in 1225, joined the then newly formed Dominican order in 1243, despite his mother's attempts to prevent him (she locked him up in the family castle for more than a year). His prime years of input to philosophy and the teachings of the Church were the 1250s and 1260s, when he managed to overcome the apparent conflict between Augustine's dependence on spiritual interpretation, and the newly re-emerging views of Aristotle, flavoured by the intermediary work of the Arab scholar Averroës, which placed much more emphasis on deductions made from the senses.

Aquinas managed to bring together these two, apparently incompatible views by suggesting that, though we can only know of things through the senses, interpretation has to come from the intellect, which is inevitably influenced by the spiritual. When considering the infinite, Aquinas put forward the interesting challenge that

> although God's power is unlimited, he still cannot make an absolutely unlimited thing, no more than he can make an unmade thing (for this involves contradictory statements being both true).[36]

Sadly, Aquinas's argument is not very useful, because it relies on the definition of a 'thing' as being inherently limited (echoing Aristotle's argument that there can't be an infinite body as a body has to be bounded by a surface, and infinity can't be totally bounded). Simply saying that 'a thing can't be infinite because a thing has to be finite' is a circular argument that doesn't take the point any further. He does, however, have another go at showing how creation can be finite, even if God is infinite, that has more logical strength.

In his book *Summa theologiae*, Aquinas also argues that nothing created can be infinite, because any set of things, whatever they might be, have to be a specific set of entities, and the way entities are specified is by numbering them off. But there are no infinite numbers, so there can be no infinite real things. This was a point of view that would have a lot going for it right through to the late nineteenth century when, as we shall see, infinite, countable sets crashed on to the mathematical scene.

The linkage of God with the infinite comes up again and again in practically all the modern world religions, from the poetry of the Hindu scriptures to Jewish mysticism. In the *Bhagavad Gita*, by tradition spoken by Krishna himself, we read:

> O Lord of the universe, I see You everywhere with infinite form . . .
> Neither do I see the beginning nor the middle nor the end of Your
> Universal Form.[37]

And in the Jewish religion, which gave birth to both the Christian and Muslim traditions, there are very specific references to infinity in the secretive collection of Jewish mystical writings known as the cabbala. This term, from which we get our word cabbalistic (though this now has darker connotations), implies the handing down of ancient wisdom.

The supporters of the cabbala – it would imply too much structure to call them a secret society – like the Pythagoreans before them were obsessed with the significance of number. But rather than approaching number from a geometrical perspective, they were particularly interested in numerology, linking each letter in the alphabet with a particular number, so words could be imbued with numerical significance.

At the heart of the cabbala were ten properties or components. Just as the Greeks were particularly fond of seven, ten had a major significance for the cabbalists – after all, God had delivered ten commandments to human creatures with ten fingers and ten toes. The ten components were called the *sefirot* (again, note the significance of number – *sefirot* means 'countings'). These components were all considered subsidiary to the Godhead – and that was given the name *Ein Sof* – in effect, infinity.[38]

More modern philosophical considerations have been not so much about God's grasp of infinity as our own. David Hume, the eighteenth-century philosopher, not only dismissed our ability to comprehend infinity, but also drew from that limitation an assumption that the infinite and particularly the infinitesimally small could not exist in reality (see page 219).

At first sight, this reluctance to accept that the finite human mind can cope with envisaging infinity seems reasonable. When Galileo, one of the Renaissance thinkers to consider the infinite, was dealing with the subject, he considered that it was incomprehensible to our finite intellect, surpassing the capacity of our imagination.[39] Yet there are ways in which it seems possible to get a handle on the infinite. When we represent a sequence like 1, 2, 3, 4, . . . , it doesn't seem so hard to envisage the idea that this progression goes on forever and never stops. We don't have to be able to grasp every number in the sequence, all the way to infinity, to see that, just as we don't have to be able to envisage every millimetre of ground between London and Hong Kong to be able to grasp the idea of travelling from one place to the other.

There is, in fact, as Shaughan Lavine, Associate Professor of Philosophy at the University of Arizona, points out in his *Understanding the Infinite*,[40] a very simple way that anyone can, to some degree, envisage infinity. As long as you can grasp the meaning of 'finite' and the meaning of 'not', you should have a simplistic picture of what infinity is all about.

The arguments about whether we (or God) can grasp the totality of infinity have little impact on the need to refer to this strange concept. After all, we can't strictly envisage even as small a number as a million, let alone the infinite. Being unable to envisage something doesn't detract from its usefulness. But to be able to work with the infinite, mathematicians needed some easy way of referring to it. Despite the early philosophers' insistence that numbers and symbols had power of their own, there was no consistent symbol allocated to infinity. It was not until the seventeenth century that this would happen. Yet it seemed inevitable. Vague, hand-waving conceptions of infinity might have been enough for theology, but when it came to mathematics, something more precise was needed.

6

LABELLING THE INFINITE

Our knowledge can only be finite, while our ignorance must necessarily be infinite.

Karl Popper, *Conjectures and Refutations*

SOME TIME IN THE MIDDLE OF THE THIRTEENTH CENTURY, a Franciscan friar by the name of Roger Bacon wrote a long, intriguing letter, possibly to William of Auvergne, the Bishop of Paris, or John of Basingstoke, an English scholar of Greek literature. In this *De mirabile potestate artis et naturae* (On the marvellous power of human workmanship and nature), Bacon points out the difference between the imaginary capabilities of magic and the very real power of natural science.

This shouldn't be too surprising – Bacon made significant steps in the development of the scientific method, emphasizing in his *Opus majus* the essential role of experiment.[41] Educated at the new universities of Oxford and Paris, Bacon had a good understanding of the accepted philosophy of the day, but was also uniquely able to put his own stamp on the need for experiment to verify philosophical hypotheses.

In his letter, Bacon makes many remarkable statements. He predicts everything from flying machines to submarines. He describes the manufacture of gunpowder. And he explores the folly that makes the ritual of magic seem capable of producing the actual results of nature. But one of his statements rings true to anyone faced with understanding mathematics:

> What beliefs ought to be held about magic symbols and characters and about similar things is the next matter for my consideration. For I doubt very much whether all things of this complexion are now false and dubious, for certain of these irrational inscriptions have been written by philosophers in their works about Nature and about Art for the purpose of hiding a secret from the unworthy...[42]

Although he had poured scorn on those who were convinced by frauds that magic could achieve anything, Bacon felt that what might have appeared to be magical incantations and inscriptions were often the result of philosophers using special symbols and characters to hide their secrets from the common herd because (as he quotes from Aristotle)

> it is stupid to offer lettuces to an ass since he is content with thistles.[43]

The mathematical fraternity would deny this entirely. As far as they are concerned they simply use necessary jargon and shorthand to be able to convey the information they need to get across in a practical fashion. The effect, though, can be just as confusing as if it were intended to conceal. In my dictionary, *jargon* is not only described as a form of words using unfamiliar or specialist terms restricted to a particular occupation or specialism. Its older meaning is unintelligible speech or writing – a collection of gibberish and nonsense. It was also used as an alternative term for a code or cipher.

The effect can be to hide the truth from those who aren't part of the club. Galileo is forced to spring to the defence of jargon in his *Dialogues Concerning Two New Sciences*, where having rather clumsily referred to two three-dimensional shapes he comments parenthetically:

> Note by the way the nature of mathematical definitions which consist merely in the imposition of names or, if you prefer, abbreviations of speech established and introduced in order to avoid the tedious

drudgery which you and I now experience simply because we have not agreed to call this surface a 'circular band' and that sharp solid portion of the bowl a 'round razor'.[44]

Yet it is hard not to subscribe to part of Bacon's view elsewhere in the letter when he says that some teachers use a particular set of symbols in order to tie the listener in to their school. If you use my symbols, you are one of my pupils and will carry forward my teaching into the world. This still happens, particularly in a new area of science before standards are established (or in pseudo-science, where the particular teacher can make up his or her own terms to taste).

Even if there is no deliberate obscurity on the part of the mathematician, the terminology often will act as a barrier, especially when used lazily – when the mathematician knows what he (or she) means, and either forgets that the common herd don't use such terminology, or can't be bothered to explain it to the rest of us.

Considering how fundamental numbers are to us in everyday use as much as in heavy-duty mathematics, very little mathematical terminology has escaped into the real world. Where much of the jargon of the computer professional has become part of normal speech, even simple mathematical terms and symbols have tended to stay as well protected as Aristotle's lettuces.

The numbers themselves we tend not to think about too much: we just use them. Yet the symbol equivalents of the words one, two, three, and so on – 1, 2, 3, . . . – had to come from somewhere. The characters we use for the numbers arrived here from India via the Arabic world. The Brahmi numerals that have been found in caves and on coins around Bombay from around the first century AD use horizontal lines for 1 to 3, like the Roman numerals but lying on their side. The squiggles used for 4 to 9, however, are clear ancestors of the numbers we use today. These symbols were gradually taken up by Arab scientists, and came to Western attention in the thirteenth century thanks to two books, one written by a traveller from Pisa, the other by a philosopher in Baghdad.

The earlier book was written by the philosopher al-Khwarizmi in the ninth century. The Latin translation *Algoritmi de numero Indorum* (al-Khwarizmi on the numbers of the Hindus),[45] which is all we now have, was produced 300 years later, and is known to be considerably changed from the original. This rather poor Latin version of al-Khwarizmi's name is generally assumed to give us the mathematical term 'algorithm', meaning a series of rules for solving a mathematical problem, though at least one significant dictionary gives the origin of the term as a derivation from the Greek *arithmos*, meaning 'number'.[46]

The translation of *De numero Indorum* slightly predates the man who is credited with introducing the system to the West, Leonardo of Pisa, who is generally known by the nickname Fibonacci (his father was Guglielmo Bonacci, and Fibonacci is a contraction of filius Bonacci, son of Bonacci). He was born around 1170 and travelled widely around North Africa and the surrounding area with his father, a Pisan diplomat. In doing so he comments in the book *Liber abaci*, written in 1202, that he was 'was introduced to the art of the Indian's nine symbols' and it was this book that really brought the Arab–Hindu system to the West.[47]

In many ways, the symbols we use are not as important as the other aspect of the numbers we inherit from India, the significance of position. Look at the Greek numbers discussed on page 22, or consider for a moment the more familiar Roman numerals. Apart from the use of a predecessor to 'chop part off' a number, so by putting the figure I (for 1) in front of X (10) we get IX (9), position has no significance in the Roman number system. This means that adjacent numbers can vary hugely in appearance – as when the year went from 1999 to 2000, shifting from to MCMXCIX to MM. And the approach makes any arithmetic that can't be done in the head a nightmare, as it isn't possible to do the simple manipulation in columns that we are familiar with, because unlike the Romans and the Greeks we have a number system that *does* use position.

To be able to use columns in this way needed a placeholder to mark a column with nothing in it – zero – the lack of which made it difficult ever to get far with the arithmetical development of mathematics in Greek and Roman times. It is a chicken-and-egg problem, whether the lack of zero forced classical mathematicians to rely mostly on a geometric approach to mathematics, or the geometric approach made 0 a meaningless concept, but the two remain strongly tied together.

As well as the simple numbers themselves, when we get into the more astronomical reaches of digits, it makes sense to take on the mathematicians' and scientists' convention of bringing back the ancient Greek superscript notation – the small number written above and to the right of another number – but now, instead of being a multiplier as in Greek times, it means that the preceding number is 'multiplied by itself this many times'. This extends the simple 2 as square (so 3^2, 3 squared, is 9) to any number. As we tend to count in tens, the most common usage is to apply this 'power' approach to 10 (power being the value of the little number, so 10^3 is 10 to the third power), where the little number simply becomes the number of noughts after the 10. Very quickly this becomes more manageable than a fully written out number. A million is to 10^6, while even the immense googol

$$10,000,000,000,000,000,000,000,000,000,000,$$
$$000,000,000,000,000,000,000,000,000,000,$$
$$000,000,000,000,000,000,000,000,000,000$$

is written compactly as 10^{100}.

The numbers we have seen so far are all simple counting numbers. Like the ratio of one number to another, they can be identified directly with real objects. The power of this number-to-object identification was the basis for the way we have already seen that the Greek school of Pythagoras made number an inherent part of creation. But the Pythagorean obsession with clean, whole numbers was to prove a real problem for them.

That handy Pythagoras' theorem about the length of the sides

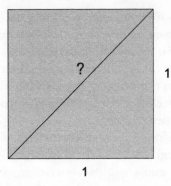

Figure 6.1 How long is the diagonal of a 1 by 1 square?

of a right-angled triangle produces a result that is frankly devastating if you believe that the universe is driven by pure whole numbers. Imagine a simple square, each side 1 cm in length. Draw a diagonal across the middle. We've now got a pair of right-angled triangles, on which we can deploy Pythagoras' theorem (see Figure 6.1).

By that theorem, we can easily work out the square of the length of this diagonal from the sum of the square of the two sides. The square of the diagonal's length is $1^2 + 1^2$, or 2, since 1^2 is simply $1 \times 1 = 1$. But what is the length of the diagonal itself? That's $\sqrt{2}$, the square root of 2 – the number which when multiplied by itself makes 2.

So far, so good, but what is that number? It clearly isn't a whole number, because $1 \times 1 = 1$ and $2 \times 2 = 4$. We are looking for something in between 1 and 2. So, could it be a fraction, or rather the ratio of two other numbers (as we have seen that fractions in the modern sense would be meaningless to the ancient Greeks)? The horrified Pythagoreans managed to prove that it was not this either.

The form they would have used to undertake this proof is rather messy, as they had no concept of using symbols to stand in for values. We can do this and see quite simply that $\sqrt{2}$ isn't a fraction.

If $\sqrt{2}$ were a fraction, then something divided by something else would be equal to the square root of 2. In case, like me, your mind always switches off when faced with x's and y's, I'm going to call the top part of the fraction '**top**' and the bottom part '**bottom**', so if it were true that $\sqrt{2}$ could be turned into a fraction it would be $\frac{\textbf{top}}{\textbf{bottom}}$.

You can skip over this description, but it does show that it is a relatively simple thing to establish that the square root of 2 can't be the ratio of two whole numbers. If you work through each step and make sure you're happy with it, you will find that the logic progresses quite painlessly.

To keep things simple, we are assuming that **top** and **bottom** provide the simplest fraction you can get – there's nothing more to cancel out, so they're like $\frac{1}{2}$ rather than $\frac{2}{4}$.

If the square root of 2 is a ratio, then we're saying that

$$\frac{\textbf{top}}{\textbf{bottom}} = \sqrt{2}.$$

Square roots are a bit fiddly, so let's multiply each side of the equals sign by itself. This gives us

$$\frac{\textbf{top}}{\textbf{bottom}} \times \frac{\textbf{top}}{\textbf{bottom}} = \sqrt{2} \times \sqrt{2}.$$

Hence,

$$\frac{\textbf{top}^2}{\textbf{bottom}^2} = 2.$$

In traditional mathematical fashion, we can get rid of the division as well by multiplying both sides of the equation by **bottom**^{2}, to get

$$\frac{\textbf{top}^2}{\textbf{bottom}^2} \times \textbf{bottom}^2 = 2 \times \textbf{bottom}^2,$$

or

$$\textbf{top}^2 = 2 \times \textbf{bottom}^2.$$

Now, the Greeks relied on their knowledge of odd and even

numbers, a subject that was afforded a lot of thought as they were supposed to represent male and female principles. They knew three things about odd and even numbers.

1. A number that can be divided by 2 is even.
2. If you multiply an odd number by an odd number, you get another odd number.
3. If you multiply any number (odd or even) by an even number, you get an even number.

As the right-hand side of the equals sign is $2 \times \textbf{bottom}^2$, the result must be even – it's the outcome of multiplying by an even number, 2. So $\textbf{top}^2$ must be even. And that means **top** has to be even (because were it odd we would be multiplying two odd numbers together and would get an odd result).

Now here comes the twist. If **top** *is* even, then it can be divided by 2. So $\textbf{top}^2$ can be divided by 4. And we know that $\textbf{top}^2$ is the same as $2 \times \textbf{bottom}^2$. If $2 \times \textbf{bottom}^2$ can be divided by 4, then $\textbf{bottom}^2$ can be divided by 2. So $\textbf{bottom}^2$ (and hence **bottom**) is even. (Read that again if necessary – it makes sense.)

So we have shown that both **top** and **bottom** are even. But if both are even, then $\frac{\textbf{top}}{\textbf{bottom}}$ isn't the simplest fraction we could have – since we can divide both **top** and **bottom** by 2. Yet we started off by saying that $\frac{\textbf{top}}{\textbf{bottom}}$ *was* the simplest fraction we could have. We've reached an impossible contradictory situation – which means our original assumption that it was possible to represent $\sqrt{2}$ by a fraction was false.

Instead, the square root of 2, the diagonal of that square with unit sides, proved to be a number which if written out as a decimal fraction just goes on and on forever, never settling into a steady repeating pattern that would mean it could be turned into a ratio. This 'forever' was the source of the nightmare. Here was a number that could not be avoided from simple geometry, but that had no place in the neat, ordered world of real numbers. As the Pythagorean picture of reality was based on number, it must have seemed that this abomination was the work of some evil spirit.

Numbers like $\sqrt{2}$ are now called irrational, not because they don't make sense, but because they can't be made out of a ratio of other numbers. There is no way to represent them as a fraction. The first inclination of the Pythagoreans was to keep the irrational numbers to themselves. They were by nature a secretive society, but it is not obvious whether this particular attempt at concealment was to keep such a fascinating discovery from the world, or to preserve the shaky foundations of their numerological philosophy. Either way, they failed.

The news got out. Legend allocates responsibility for this early example of a news leak to one Hippasus of Metapontium, who may also have made the discovery himself. According to one version of the legend he was expelled from the school. A more dramatic version has Hippasus sharing his discovery with his colleagues (or being found out as the source of the leak) while on a boat trip. Such was the concern of the Pythagoreans to keep irrational numbers to themselves that Hippasus was thrown over the side of the boat to drown for daring to threaten the purity of number.[48]

Bearing in mind the complexities of Greek mathematical notation (see Chapter 3), it might seem amazing that the Pythagoreans could cope with the values involved in this discovery. After all, as the square root of 2 can't be represented as a ratio, we have to use a decimal fraction to write it out. Indeed it would be amazing, were it true that the Greeks did have a grasp for the existence of irrational numbers as a fraction. In fact, though you will find it mentioned in many books that this was what the Pythagoreans did, to talk about them understanding numbers in this way totally misrepresents the way they thought.

As we have seen, the ancient Greek mathematicians did not perform mathematical manipulations with numbers and x's and all the other tools of their later colleagues. Their observations were geometrical – they worked not with formulae but with diagrams (hence all those tedious theorems anyone who has suffered traditional geometry had to go through). In performing these geometrical feats they weren't using ordinary rational fractions – instead they thought of the proportion between two lengths – they might know, for instance, that one side of a triangle was twice the size of another.

When dealing with such ratios, they would know that there was a clear relationship in terms of a full unit – so, for instance, in the famous right-angled triangle of Pythagoras' theorem, they would think of the longest side being 5 units long when the other sides were 3 and 4 – but not of the smallest side being $\frac{3}{5}$ of the length of the longest. What they then discovered about the square root of 2, thanks to the triangle with smaller sides of 1 and 1, is that the length of the longest side, which we would now think of as $\sqrt{2}$, was not measurable.

They did not have the concept of an irrational fraction – instead they would say that the length of the side had a value greater than 1 and less than 2 that could not be measured. In fact there have been suggestions that the whole business of the Pythagoreans being horrified by this non-number is a romantic fiction dreamed up after the event. Because they entirely separated their theory of number (and the perfection of number) from geometry, which was performed in a purely visual manner, there was no corruption of the perfection that number implied. Irrationals simply didn't exist as numbers.

By the middle ages, though, the existence of the irrational numbers was grudgingly recognized, even though they were still treated with deep suspicion. Take the view of Michael Stifel, the sixteenth-century German mathematician, who is best known for inventing logarithms independently of the 'true' inventor Napier. Stifel pointed out that irrational numbers were valuable because they could explain things that couldn't be managed with rational numbers alone. However,

> on the other hand, other considerations compel us to deny that irrational numbers are numbers at all. To wit, when we seek [to represent

them as decimals] . . . we find that they flee away perpetually, so that not one of them can be apprehended precisely in itself.[49]

Stifel grumbles that irrationals are not true numbers but lie 'hidden in a kind of cloud of infinity'.

Yet the irrational that had cropped up in the simple diagonal of a unit square is no real problem to work with. You might not be able to actually write it down in full, but it is very easy to say what it is: $\sqrt{2}$, the number which when multiplied by itself gives you 2. We can write a simple formula for it: if we call it 'roottwo', then

$$\text{roottwo} \times \text{roottwo} = \text{roottwo}^2 = 2.$$

It's not exactly fearsome. But other irrational numbers are much less amenable to simple understanding. Such numbers share the property of being impossible to write out in full, as they never fall into a repeating pattern, but they won't even fit to a nice, simple formula, a way of getting a mental handle on what they are. It might seem strange, then, that the best-known such number was also familiar in antiquity, but once again we are being misled by our fixation with the symbols and structures of algebra. This particular irrational number may not be easy to put into an equation, but it is trivially represented by a diagram such as that shown in Figure 6.2.

How much bigger is the circumference of a circle than its diameter? What is the ratio of one to the other? Not surprisingly, with their graphical view of mathematics, this challenge fascinated the Greeks. There was no obvious way to make it a real ratio, a ratio of whole numbers. There was something intractable about the circle. Even the apparently simple problem of finding a square with the same area as a circle didn't work easily.

Thinking visually in the Greek fashion, this seems a natural enough thing to be able to do. You can imagine treating the circumference of the square as if it were a rubber band, gradually increasing its area until in reached the exact value of the circle's area. Such was the fascination of this apparently simple puzzle

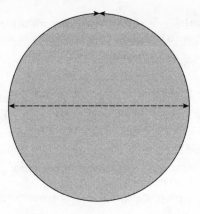

Figure 6.2 A strange irrational.

that the Greeks even had a word, τετραγονίζειν (*tetragonizein*), for attempting to translate the area of a circle into a square. The challenge, known as 'squaring the circle' (technically using only a ruler and dividers), wasted thousands of hours of philosophers' and mathematicians' time. But the devil is in the detail, and while it is possible to match the areas to a certain degree, the problem arises when pinning down the exact area.

The reason behind this is its dependence on an irrational number, one that unlike $\sqrt{2}$ has no simple relationship to whole numbers – that number, of course, that we now call pi (π). As the number is irrational we can never write it down completely, and as it doesn't have a relationship to whole numbers – you can't put together an equation that derives π or a power of it from a finite set of other numbers or powers – it seems inevitable now that it is impossible to match the area of a circle exactly to that of a square. These two steps in laying the problem to rest took a long time to take. It wasn't until 1768 that the Swiss mathematician Johann Lambert showed that π was irrational, and as late as 1882 that the German C. L. Lindemann proved it had no value that could be defined using an equation of squares or other powers, and hence that squaring the circle was impossible.

To the Greeks this was simply the ratio of the circumference of a circle to its diameter, a ratio without a measurable value. One that stubbornly refused to become related to the sides of a square, as the diagonal does to the sides of a triangle in $\sqrt{2}$. Whereas $\sqrt{2}$, and every other irrational number that can be defined with an equation, is called algebraic to echo this property, π is far and above the best known transcendental number, the name given to irrationals that can't be fitted to a suitable finite equation. Just as irrational does not suggest lacking rational thought, transcendental has nothing to do with the mystical associations the word has picked up in the last few years. It merely says that the number transcends – is outside of – calculation by equation.

Although this is the case with π, we now know the value of this quirky number to many thousands of decimal places. Yet it's hard enough to remember it, let alone calculate it. When I was at school we made do with 3.14159 – or whatever the calculator said when you pressed the π button. Others have resorted to rhymes to fix the sequence of numbers in memory, using words of the right number of letters for each value in turn. In 1906 one Adam C. Orr of Chicago sent the following verse to the *Literary Digest*, which carries π to a fair number of decimal places:

> Now I – even I – would celebrate
> In rhymes unapt the great
> Immortal Syracusan rivalled nevermore,
> Who in his wondrous lore,
> Passed on before,
> Left men his guidance
> How to circles mensurate.[50]

When I first heard about attempts to add more and more decimal places to π, I imagined the mathematicians who were achieving this feat using more and more accurate rulers and protractors to pin down the ratio of a circle's circumference to its diameter. In fact, though, Archimedes had made a reasonable go at it by working out the area of a 96-sided polygon fitted inside the circle, and though π cannot be worked out from a finite equation, it can be calculated using an infinite series – a rather more sophisticated version of

$$1 + \tfrac{1}{2} + \tfrac{1}{4} + \tfrac{1}{8} + \tfrac{1}{16} + \tfrac{1}{32} + \cdots,$$

but in this case a series that instead of heading for 2, homes in on the value of π. There are several of these, the earliest of which dates back to the late sixteenth century, but the first simple formula was discovered by John Wallis (see below) who was able to show that

$$\tfrac{1}{2}\pi = \tfrac{2}{1} \times \tfrac{2}{3} \times \tfrac{4}{3} \times \tfrac{4}{5} \times \tfrac{6}{5} \times \cdots,$$

where the top part of each fraction goes up by two in pairs, beginning at 2, and the bottom part of each fraction goes up by two in pairs after the initial value of 1. Because an infinite series (or, in this case, product) never quite makes it to the 'true' value, we can't specify π exactly this way, but given enough time, it is possible to pin down π to as many decimal places as you would like.

Numbers like π that refuse ever to fall into a pattern, continuing forever in a way that can be worked out but never predicted, made it impossible to totally ignore infinity itself, even if Aristotle had decided that it was enough to have potential infinity. Just as other aspects of mathematics were settling down to a standard set of symbols, it became necessary to give infinity a convenient label.

Where did that collapsed figure-eight symbol ∞ come from? Exactly *why* it was used isn't clear, but it first appeared in the work of the English mathematician John Wallis. In *De sectionibus conicis* (On conic sections), written in 1655, Wallis explored the subject of the interesting, but hardly mind-stretching, concept of conic sections – the shapes produced when different cuts are made through a pair of cones joined at their apexes, and threw in the symbol ∞ almost in passing, as if were already commonplace. It appeared again in the more substantial *Arithmetica infinitorum*, which he wrote in 1656. Here, Wallis explored infinite series and covered some aspects of calculus, at the time the newest aspect of mathematics to be based around the concept of infinity.

Wallis is the great forgotten mathematician of the seventeenth century (at least, forgotten outside of mathematical circles), bleached out of the public gaze by the incandescent brilliance of Newton's fame and genius. John Wallis was ideally placed to consider infinity. Although he had become Savilian professor of geometry at Oxford by the time he wrote his book, he had originally studied theology, combining well the more mystical and practical aspects of the infinite. It is possible that the disappearance of Wallis from public view is simply because his name was swamped by those of famous contemporaries. This was not just the era of Newton, but also of Descartes, Fermat, Halley, Pascal and Hooke. Even so, it is strange that Wallis should not have captured more of the public imagination. Like the names that have better survived, Wallace was a polymath, interested in a wide range of subjects and fascinated not only by mathematics but also by theology and physics.

It was in 1616, on 23 November, that Wallis was born at Ashford in Kent. Now the site of a major international railway station for the Channel Tunnel, at the time Ashford was a backwater country town. One key factor in John Wallis's early life is reminiscent of that of Isaac Newton, born 26 years later. Both boys lost their fathers when they were young – in Wallis's case, he was six. But where Newton's was a farming family, putting pressure on him to stay away from academia and keep the farm going, Wallis senior had been rector of Ashford. His family encouraged the boy's education, and young Wallis was sent to a series of schools, ending up at Emmanuel College, Cambridge, in 1632.

At this stage, any knowledge Wallis had gained in mathematics was no more than a hobby. It was not a subject that was taken too seriously in the schools of the day. When he went up to Cambridge, his education would centre on the classics, rhetoric and logic. Although the scientist-friar Roger Bacon had emphasized the importance of mathematics back in the 1260s, writing 'He who is ignorant of mathematics cannot know the other sciences and the things of this world',[51] this view had not become a popular one even by Wallis's time.

Bacon's Elizabethan namesake, Francis Bacon, though prepared to develop intricate methodologies for science, regarded mathematics as of little import in the sciences. Interestingly, Wallis would later pick up on the remarkably advanced nature of Roger Bacon's thinking, writing to the great German mathematician Gottfried Wilhelm von Leibniz that Bacon's joining of mathematics to natural philosophy, which was essential to the advancement of physics, was some 400 years ahead of his time.[52]

While at Cambridge, though, Wallis's career seemed to be heading in a different direction. Wallis maintained his interest in mathematics, but found it difficult to get together with others of a similar mind. He later commented ruefully that, of the more than 200 students in his college, he could not find two with a greater knowledge of mathematics than his own, poor though this was. Having said that, there does seem to have been a fair amount of mathematical activity at the universities by then – perhaps Wallis was unlucky in his choice of college.

Wallis gained his BA and subsequently his MA at Cambridge, and was ordained in the same year, 1640. By now it was obvious that the young man had great potential, and an attempt was made to make him a Fellow of the college, but this was thwarted by a bizarre rule. It seems that at the time there could only be a single Fellow from each county. As a Mr Wellar already represented Kent, Wallis had to be excluded. Instead he became a Fellow of Queens' College – but his position was not to last long. Within months he took up a position as chaplain to a wealthy Yorkshire landowner.

Initially Wallis's intention was that the only academic work he would continue with would be in theology – divinity, as it is still called at Cambridge – but the realities of warfare imposed a new interest. The Civil War, which began in 1642, brought a new necessity to the already strong fascination of codes and ciphers. (Technically a code replaces words with another word or character, while ciphers work on a letter-by-letter basis.)

Codes had been used since ancient times to pass information secretly, but by the seventeenth century they had become a fact of everyday political and military life. Rather less than a hundred years earlier, it had been a dependence on a weak cipher that had led to the downfall of Mary, Queen of Scots. Her trial for treason in 1586 was based on a series of letters that implicated her in a plot to murder the protestant English queen, her cousin Elizabeth.

These letters had been enciphered, but the method used was little better than a simple substitution of an artificial set of characters for the alphabet. Admittedly there were certain subtleties – a few words had their own special code letter, and there were a number of different meaningless characters, plus a special character to identify a double letter. Even so, this cipher proved susceptible to the already well-known technique of frequency analysis – looking at how often different characters appeared, and matching this frequency with the typical use of various letters of the alphabet. Mary might as well have signed a confession.

By the time of the English Civil War, cryptographic techniques had been improved, although there was still a lot of reliance on variants of simple substitution. The need to pass information between members of the Royalist and Parliamentarian armies brought a flourishing trade in the production of new ciphers. Wallis was plunged into this cloak and dagger world when he came across a secret document on the subject of the capture of Chichester and tried to break it as an intellectual challenge. He proved a natural, with the pattern-spotting eye of the crossword enthusiast.

Wallis went on to become a cryptographic expert on behalf of the Parliamentarians, decoding any Royalist messages that fell into their hands with remarkable ease. The fascinating puzzle of untangling codes developed into a passion for Wallis. At about the same time he began his association with the newly founded Royal Society, a meeting place for discussing the wonders of nature from the vivisection of crocodiles and the existence of werewolves to the nature of light and the fundamentals of mathematics (all topics discussed in the early years). The Society brought more of the tantalizing challenges of mathematical theory to Wallis's grasp.

He picked up the current hot mathematical topics with remarkable speed, but arguably Wallis was still an enthusiastic amateur when he was awarded the Savilian chair in geometry at Oxford in 1649. His sudden ascension to the dizzy heights of the mathematical hierarchy is less surprising when it is realized that it was Oliver Cromwell who put him there – the same Oliver Cromwell who had benefited so well from Wallis's cryptographic services to the Parliamentarians. But there is no doubt that Wallis deserved the position, and was able to hold on to it after the restoration of the monarchy in the early 1660s. In fact he was to keep the chair for 50 years.

It is in his use of indivisibles (a term we will return to in Chapter 8), tiny fractional divisions of an object, that the symbol ∞ for infinity first appears in Wallis's work. Imagining a plane to be made up of an infinite number of extremely narrow rectangles (or to be precise, parallelograms), Wallis casually introduces the symbol: he might just as easily have used the letter 'i' for infinite, but instead he wrote:

> Let the altitude of each one of these [parallelograms] be an infinitely small part, $1/\infty$, of the whole altitude and let the symbol ∞ denote Infinity.[53]

Wallis does not tell us why he chose this particular symbol. Some have suggested that the symbol is derived from the old Roman sign for 1000 'CIƆ' or from a closed-up version of the last letter of the Greek alphabet, omega (ω). Elsewhere, it has been suggested that it brings to mind the infinity of a loop – but so does a circle, or any other line you wish to draw that joins up with itself.

There is also a feeling of a Möbius strip about it – one of those twisted and joined strips of paper that have the strange property of only having a single face. Yet it is unlikely that Wallis was trying to suggest both the continuity of a loop and the unity, the singularity, of such a strip. He predates Möbius by nearly 200 years. Whatever his reasoning, we were all to follow his suggestion 'let the symbol ∞ denote Infinity'.

Are we now fluent in the language of infinite mathematics, having given infinity a label? Hardly. For example, that familiar ∞ will prove not to be a symbol for infinity at all, but something rather less grand. Even so, further distinctions can be introduced as we come to them. It's time to get away from symbols and look at the real thing – or at least, the first tentative steps towards infinity that were all that most mathematicians would risk until the end of the nineteenth century. The man who would give us a preview of the infinite is much better remembered than Wallis, although it is physics rather than mathematics that has kept him in the public eye.

7

PEEKING UNDER THE CARPET

Man's unhappiness, as I construe, comes from his greatness; it is because there is an Infinite in him, which with all his cunning he cannot quite bury under the Finite.

Thomas Carlyle, *Sartor Resartus*, Book ii, Chap. 9

FROM THE GREEKS ONWARDS, infinity was regarded with suspicion. Realizing the dangers of adding up infinite series, a trick was employed to avoid ever dealing with infinity itself. Instead of trying to work out the sum of a series that goes on forever, a finite series was used, one that had a known number of terms. So, instead of saying, for instance, that the sum of our old friend

$$1 + \tfrac{1}{2} + \tfrac{1}{4} + \tfrac{1}{8} + \tfrac{1}{16} + \tfrac{1}{32} + \cdots$$

actually *was* 2, the approach was taken that the sum of a finite part of the series can be made as close as you like to 2, without ever actually reaching it. It seemed enough that for any gap between the sum of the series and 2, you could define a number of terms in the series that would make it happen. The series is said to 'converge on 2' – it heads inevitably in that direction, even though in real terms it never quite makes it.

The great mathematician Carl Friedrich Gauss, who lived from 1777 to 1855, emphasized this underlining of the unreal nature of infinity. He wrote:

> I protest against the use of an infinite quantity as an actual entity; this is never allowed in mathematics. The infinite is only a manner of speaking [façon de parler], in which one properly speaks of limits to which certain ratios can come as near as desired, while others are permitted to increase without bound.[54]

In other words, infinity is just an illusion. Like the end of the rainbow, you can never actually get there, but it's something very handy to aim for. As we shall see in Chapter 9, the language of calculus is full of evasions like 'the value tends to 10 as x tends to infinity'. It's a sneaky way of saying that the value of something becomes 10 when x is infinity (where x is some value in an equation), but without ever needing to admit that infinity exists.

This sweeping under the carpet of infinity that had begun with Aristotle, representing it as a potential concept that never comes into being, would remain the only acceptable view of the concept for most scientists and mathematicians from the ancient Greeks through to the late nineteenth century, but this did not mean that everyone was prepared to leave infinity tucked away. Some individuals could not resist a peek under the carpet, and one of these was the remarkable Galileo Galilei.

To get an insight into Galileo's views on infinity, we have to fast forward through a life that is already well enough documented elsewhere. Born in 1564 into the family of a musician with an amateur bent for the scientific, Galileo grew up to combine a razor-sharp insight with an eye to the main chance that is illustrated well by the way that he managed to intercept a Dutch telescope builder on the way to Venice to sell his new product. While a friend kept the unfortunate optician busy, Galileo managed to build his own telescope. With the Dutchman still held up, Galileo was then able to bring his device first to the Venetian senate, which had unwittingly played into Galileo's hands by asking his friend to look into the new technology, giving him advance warning of the opportunity.

It was only after the publication in 1632 of Galileo's famous book supporting the Copernican theory that put the Sun and not the Earth at the centre of things, *Dialogue on the Two Principal World Systems*, that he was to commit his ideas on infinity to paper. By then he had been brought before the Inquisition, had been tried and imprisoned. It was *Dialogue on the Two Principal World Systems* that caused his precipitous fall from favour. Although Galileo had been careful to have his book passed by the official censor, it still fell foul of the religious authorities, particularly as Galileo had put into the mouth of his 'dim but traditional' character Simplicio an afterword that could be taken to be the viewpoint of the Pope. This seemed to imply that the pontiff himself was backward in his thinking.

Whether triggered by this apparent disrespect, or the antipathy a man of Galileo's ebullient character would inevitably generate in a bureaucracy, the authorities decided he needed to be taught a lesson. Someone dug back in the records and found that Galileo had been warned off this particular astronomical topic before. When he first mentioned the Copernican theory in writing, back in 1616, it had been decided that putting the Sun at the centre of the universe rather than the Earth was nothing short of heretical. Galileo had been told that he must not hold or defend such views.

Crucially, though, the Inquisition had also been given at the time the option to forbid him from holding, defending or *teaching* such views if he would not agree to the restriction. There is no evidence that this third part of the injunction was ever put in place. The distinction here is that Galileo should have been allowed to teach (and write about) the idea of a Sun-centred universe provided he did not try to show that it was actually true. Although there is no record that Galileo went against this instruction, the Inquisition acted as if he had.

This enabled the Inquisition to support its trial of Galileo for heresy with an apparent breach of a ruling that should never have been issued. Although Galileo had documentation showing that he had only ever been officially warned off holding and defending the views, views he claimed that he had long ago dismissed, the Inquisition was not to be fobbed off. However, Galileo had powerful defenders who managed to commute what could have been a capital sentence to house arrest for life.

Before his final confinement at his home in Arcetri, Galileo spent a number of months under the watchful eye of the Archbishop of Siena (a sympathizer, though not prepared to argue against the Inquisition on Galileo's behalf). The five months Galileo spent in Siena gave him an opportunity to pull together various strands of thinking that he had been working on for many years – while still in Siena he wrote that he had completed 'a treatise on the new branch of mechanics full of interesting and useful ideas'. By the time he had settled in at Arcetri, Galileo was able to send a message to his former pupil, Father Fulgentius Micanzio that 'the work was ready'.

The work in question was *Discourses and Mathematical Demonstrations Concerning Two New Sciences*, a book that is generally considered Galileo's masterpiece, his equivalent of Newton's *Principia*. As if to demonstrate finally and without doubt that he had done nothing wrong in the way he had structured his banned work on the Solar System, Galileo wrote his new book in exactly the same format – as a discussion between three characters, Filipo Salviati (named after a friend of Galileo's who died 16 years before in 1614), representing Galileo's 'new' viewpoints, Simplicio, who remained stuck firmly in the ancient Greek tradition, and Giovanfrancesco Sagredo (named for another friend, who had died in 1620). Sagredo provided the viewpoint of an independent observer, listening to the discussion and pulling out points to highlight.

With his masterpiece assembled, the old Galileo would have rushed into publication, making sure that a rich patron's name was splashed widely across the front. As it was, he was in a different world.

Now Galileo had the weight of the Holy Office of the Inquisition hanging over him. The Italian publishing houses in Florence and Rome would not touch anything he had written. He asked his friend Father Micanzio if he could test the waters in Venice for him – a publisher there had rescued his near-fatal *Dialogue on the Two Principal World Systems* when the others had turned their back on it. Micanzio spread the word in Venice and got enough interest to ask Galileo for a sample. Venice was always enthusiastic to demonstrate its independence from the power structures of Rome. However, Micanzio had no intention of going against the Holy Office and asked the Inquisition for an informal assessment of the chances of Galileo being allowed to publish. The response had all the subtlety of a razor blade. He was served with an express order prohibiting the printing or reprinting of anything Galileo produced. It did not matter if it was in Venice or anywhere else. There would be no exceptions.

Galileo was not so easily discouraged. He began to work his contacts to find a route to get his new book published in a way that would not bring him back under the baleful eye of the Inquisitor. He toyed with the possibility of publishing in Germany, in Vienna, in Prague – yet even this far from Rome there was a danger of a backlash. In the end, it was the visit of Dutch publisher Louis (Lodewijk) Elzevir to Italy in 1636 that seemed to open an ideal pathway to a safe, Protestant printing. Elzevir, the third Louis in that great publishing family, took an incomplete manuscript back to Leyden.

The final book, printed in 1638, was not quite what Galileo had hoped for. As is so often the case, for example, the author felt irritated that the publishers had changed his title. We don't know what Galileo had intended it to be, but he called the printed title, in full translating as *Discourses and Mathematical Demonstrations Concerning Two New Sciences pertaining to Mechanics and Local Motions*, that the publishers had substituted 'a low and common title for the noble and dignified one carried upon the title page'.[55]

While we might have sympathy with Galileo for the way his title was changed, it is also possible to acknowledge the frustration that Elzevir seemed to be feeling during the process of getting the book together as, like so many other authors, Galileo continued to try to make changes until much too close to publication. The dialogues are divided into four 'days', but Galileo continued to refer in letters to Elzevir to a fifth day on 'the force of percussion and the use of catenary'. Though Galileo comments that he had 'almost reached a complete solution' he had not got the manuscript in place in time, and the publication went ahead without it.

The final consideration, always so important to Galileo, was the dedication of the book. He wanted to make it out to the Count François de Noailles. This one-time soldier, who had been one of Galileo's pupils at Padua, was now the French ambassador in Rome. Noailles had done his best to minimize the impact of the trial on Galileo, and he seemed an ideal recipient of Galileo's gratitude – but given the outright ban on publication from the Inquisition, Galileo had to find a form of words that would protect both himself and Noailles.

The approach he decided to take would seem to be too naive to fool the Inquisitors, but perhaps they decided on leniency. In his dedication, Galileo seems to be downright surprised this book ever was published. He tells the count:

> Indeed, I had decided not to publish any more of my work. And yet in order to save it from complete oblivion, it seemed to me wise to leave a manuscript copy in some place where it would be available at least to those who follow intelligently the subjects which I have treated. Accordingly I chose first to place my work in your Lordship's hands, asking no more worthy depositary...
>
> A little later, just as I was on the point of sending other copies to Germany, Flanders, England, Spain and possibly to some places in Italy, I was notified by the Elzevirs that they had these works of mine in press and that I ought to decide upon a dedication and send them a reply at once. This sudden and unexpected news led me to think that the eagerness of your Lordship to revive and spread my name by passing these works on to various friends was the real cause their falling into the hands of printers who, because they

had already published other works of mine, now wished to honour me with a beautiful and ornate edition of this work.[56]

Sudden and unexpected news? Hardly. And the attempt to divert attention from Noailles by suggesting that he had only shown his works to various friends who were 'the real cause for their falling into the hands of the printers' is clumsy in the extreme. Yet Galileo was allowed to get away with it, and the *Dialogues* was published without difficulty.

Unlike the news that the book was to be printed, one thing truly is sudden and unexpected – the way that the topic of infinity comes into the *Dialogues*. The three protagonists are trying to work out what it is that holds matter together. Salviati reckons it is down to the vacuum that exists between the tiny particles of matter, as everyone knows that a vacuum between two objects makes it hard to separate them. This, he believes is why a metal loses its solidity in a fire, because the fire gets in between the particles filling up the gaps, and there is no more vacuum to hold them together. Simplicio is scornful. After all, the amount of vacuum must be tiny – and hence there must only be a tiny force holding matter together, not exactly what is found if you try to break apart a piece of metal or stone by hand.

It is this misguided but clever attempt to come up with an explanation of the nature of solid matter that leads us into some quite remarkable contemplation of infinity. Sagredo comes to Salviati's aid with a rather neat illustration of the way small forces can add up to something significant:

> There is no doubt that any resistance, as long as it is not infinite, may be overcome by a multitude of minute forces. Thus a vast number of ants might carry ashore a ship laden with grain . . . If you take another number four or six times as great, and if you set to work a corresponding number of ants they will carry the grain ashore and the boat also. It is true that this will call for a prodigious number of ants, but in my opinion this is precisely the case with the vacua which bind together the least particles of metal.[57]

Salviati, ever one to challenge the status quo, teases Sagredo about his proviso 'as long as it is not infinite': 'But even if this demanded an infinite number, would you still think it impossible?'

Sagredo hesitates. 'Not if the mass of metal were infinite; otherwise . . .'

Salviati goes for the kill. 'Otherwise what? Now since we have arrived at paradoxes, let us see if we cannot prove that within a finite extent it is possible to discover an infinite number of vacua.'

Although Galileo's stated intent is to see how practical the concept of an infinite number of vacua is, in practice the next 20 or so pages of the book are a clear voyage of delight around some of the mental contortions that infinity generates in us all — and establishing some truths about the nature of infinity that would keep it from being considered as a 'real' quantity for more than 200 years.

Salviati amazes his friends by asking them to consider a real, physical machine that seems to perform the impossible in a way that requires the involvement of infinity. Galileo's actual illustration used a geometrical description that baffles the eye in its number of points labelled A to Z, but the argument is simple. He

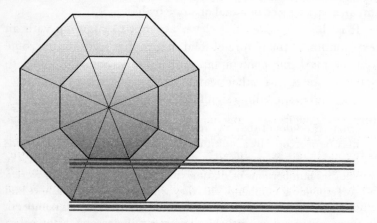

Figure 7.1   The two brass octagons.

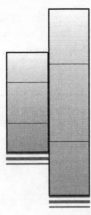

Figure 7.2 The brass octagons, end view.

first imagines a multiple-sided shape – an octagon for example. It's easier to picture as a physical object. Imagine a symmetrical brass octagon, rolling in the rather bumpy way of a non-circular wheel along a track. It doesn't have to be brass, but somehow the image is more satisfying if made out of solidly Victorian materials.

Now let's fix another octagonal wheel, around half the size, in the middle of the first wheel. Owing to a bit of cunning engineering, this also is running along a track. The easiest way to imagine this is to think that the smaller octagon sticks out of the side of the big one, so the small octagon's track is parallel to the other, but raised up and offset to one side. From the side it looks something like Figure 7.1, and from the end like Figure 7.2.

We're ready to begin. Rotate the big octagon one-eighth of a turn, so it goes from having one side flat on the track to having the next side flat down. Now check what's happened to the smaller octagon. It too has rotated by one flat side – yet somehow the whole wheel has moved on by the size of the big wheel, much further than the length of the small octagon's side. The leading edge of the small octagon's side touching the track has moved forward the whole length of the big octagon's side. How? It's simple enough – as the big octagon's side rotated it lifted the small

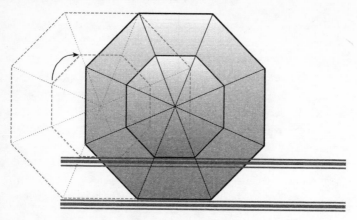

Figure 7.3 The octagonal wheel moved on.

octagon right off the track – it left a gap in the motion down the track of the small octagon equivalent to the extra distance.

So far, so good. And Salviati points out that it doesn't matter how many sides our wheel has, the same holds true. But now let's look at the same picture with a pair of circular wheels. Presumably things are much the same.

Once again we have two wheels fixed together, one smaller

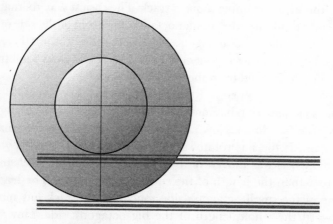

Figure 7.4 The two brass circles.

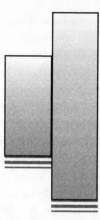

Figure 7.5 The two brass circles, end view.

than the other, each running on its own track, so that sideways on they look something like Figure 7.5.

Now let's rotate the big wheel a quarter turn. It's no surprise that the smaller wheel makes a quarter turn too. Just as with the octagon, the point at which the small wheel touches the track has now moved forward along its own track the same distance as the point where the big wheel touches *its* track.

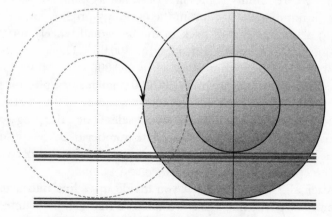

Figure 7.6 The circular wheel, moved on.

But there's a real problem here. Unlike the octagonal wheel, the circles are smooth. They are continuous pieces of metal. If we made the machine well enough, there was no time when either wheel was out of contact with the track. So imagine following the point at which each wheel touches the track as we make the quarter turn. The big wheel has measured out a quarter of its circumference along the track. And the small wheel had a quarter of its, much smaller, circumference touch its track. Yet somehow, without ever seeming to leave the track, the smaller wheel has jumped on the full distance that the big wheel travelled. How did it manage to get so much further along? Where was the gap?

If you aren't quite sure what the fuss is about, imagine that we had the wheel in its starting position and undid a quarter of the rim of both the big and small circle. Flatten it out on the tracks and you will find that the big wheel's undone rim covers the distance the whole wheel moved. But the small wheel's rim is too short. How did it manage to cover the extra ground without ever coming away from the track the way the small octagon did?

It is a worrying puzzle. As Sagredo ruefully admits: 'This is a very intricate matter. I see no solution. Pray explain it to us.'

Salviati, a.k.a. Galileo, has an explanation. This is a circumstance where infinity comes into play, not as some abstract concept in number theory, but in the everyday progress of a wheel down a track. He goes back to the octagonal wheel, but then imagines giving it not eight sides, but 100,000. Then, he says, as the big wheel traces out a line of its 100,000 sides, so the small wheel's track is made up of the 100,000 smaller sides, plus 100,000 tiny gaps between each side. Imagine this going to the extreme of an infinite number of infinitely small sides. Then the progress of the smaller circle also has an infinite number of infinitely small sides, with infinitesimal gaps in between – gaps that are, of course, imperceptible.

After all, says, Salviati, if you divide up a line into a finite number of parts, that's to say a number that can be counted, it isn't possible to arrange them into a greater length than they first occupied without there being gaps between. But with an infinite number of infinitely small parts, he imagines that we could conceive of a line extended indefinitely by an infinite number of infinitely small empty spaces.

Simplicio bridles at the thought of working anything out by a combination of infinitely large numbers of infinitely small elements, and Salviati is forced to agree – but still he feels it is possible to indulge in speculation. He says:

> These difficulties are real; and they are not the only ones. But let us remember that we are dealing with infinities and indivisibles [infinitesimals], both of which transcend our finite understanding, the former on account of their magnitude, the latter because of their smallness. In spite of this, men cannot refrain from discussing them, even though it must be done in a roundabout way.
>
> Therefore I also should like to take the liberty to present some of my ideas which, though not necessarily convincing, would, on account of their novelty, at least, prove somewhat startling. But such a diversion might carry us too far away from the subject under discussion and might therefore appear to you inopportune and not very pleasing.

Luckily for us, Sagredo encourages the digression:

> Pray let us enjoy the advantages and privileges which come from conversation between friends, especially upon subjects freely chosen and not forced upon us, a matter vastly different from dealing with dead books which give rise to many doubts but remove none. Share with us, therefore, the thoughts which our discussion has suggested to you; for since we are free from urgent business there will be abundant time to pursue the topics already mentioned; and in particular the objections raised by Simplicio ought not in any wise to be neglected.

Galileo can't resist the opportunity in his reference to 'dead books' to take a dig at those whose whole ideas of science are based on recycling the works of the ancient Greek philosophers. In demonstrating some of the wonders of infinity, he starts by

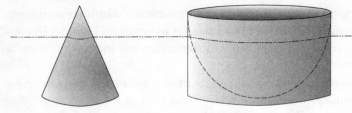

Figure 7.7 Galileo's cone and bowl.

demonstrating a geometrical proof that involves taking a cut through a pair of three-dimensional shapes (a cone, and the 'bowl' left by cutting a hemisphere out of the top of a cylindrical block).

He shows that it is possible to make the sizes of these two objects such that if you move the place you cut (the dotted line in the diagram) up or down both shapes simultaneously, they each have the same volume and the same area on the surface of the cut. Yet as we move the place the cut is made towards the point of the cone and the rim of the bowl (which are level) we end up in the one case with a point and the other a circle. He seems to have shown us a point and a line (the circumference of the circle) which have the same 'size'.

Salviati's conversational partners are impressed, but still Simplicio has a problem. It's as if all the time Salviati has been discussing geometric proofs, Simplicio has still been worrying away at the problem of the two brass wheels. If the solution depends on saying that both the longer line traced by the big circular wheel and the shorter line traced by the smaller circular wheel have an infinite number of points in them, aren't we saying that there are two types of infinity, one bigger than the other? Because surely the infinity of points on the longer line is greater than the infinity of points in a shorter one.

'This assigning to an infinite quantity,' he says, 'a value greater than infinity is quite beyond my comprehension.'

Salviati admits it is mind-boggling: 'This is one of the difficulties which arise when we attempt, with our finite minds, to discuss the infinite, assigning to it those properties which we give to the finite and limited; but this I think is wrong, for we cannot speak of infinite quantities as being the one greater or less than or equal to another.'

Galileo sets out to prove that there is no difference between the infinity of points in the two wheels by getting Salviati to ask Simplicio a series of questions that forces him into a position of accepting that apparently different-sized infinite numbers are in fact the same infinity. He checks that Simplicio knows what a square is – a number multiplied by itself – and also what roots are – the number that is being multiplied by itself to produce the square.

He then goes on to use roots and squares to prove the apparently impossible. For each root, there has to be a square. You can work through each of the counting numbers and assign each its square. For example,

$$1 \to 1, \quad 2 \to 4, \quad 3 \to 9, \quad 4 \to 16, \quad 5 \to 25,$$

and so on. Yet not every number *is* a square. In that sequence, we missed out 2, 3, 5, 6, 7, 8, and so on, from the squares. In the short list we looked at, there were 25 numbers, but only 5 squares.

Now comes the crunch. Carry on all the way to infinity. You must have a square corresponding to every single number. There are an infinite number of both roots and squares. For every whole number we can come up with an equivalent square. Yet in the list of roots there are plenty of numbers that aren't squares. Galileo has Salviati conclude that the totality of numbers is infinite and the number of squares is infinite, even though these would seem to have different values, because there are plenty of numbers that aren't squares. In fact, he suggests, the attributes 'equal', 'greater' and 'less' are not applicable to infinite but only to finite quantities.

Galileo had spotted something very special about infinity. If it can be thought of as a number at all, it is not one that obeys the normal rules of arithmetic. There could be two 'different' infinities, one clearly a subset of the other, that nonetheless are of the same size. And this made it possible for Galileo's two wheels to cover the same distance despite their differences in diameter.

But as Galileo's characters admitted to each other in their chatty fashion, dealing with an infinitely large number of infinitely small quantities was a tricky business. How far could you go? He describes these small quantities as indivisibles – what exactly did that mean? The concept would cause much confusion before paving the way for one of the most useful – and most heavily disputed – techniques ever to come out of mathematics into the real world.

8

THE INDIVISIBLE MYSTERY

My anatomy is only part of an infinitely complex organisation, my self.
Angela Carter, *The Sadeian Woman* (Polemical Preface)

THERE WAS SOMETHING IMMEDIATE, something practical, about the infinitely small segments that Galileo imagined making up the circumference of a wheel. They might be just as vague in terms of their existence as the vastness of infinity itself, but they were somehow more tangible. In the seventeenth century, when enthusiasm for these near-non-existent quantities was at its peak, they would be referred to as indivisibles, the ultimate splitting up of an item.

It would certainly be useful if indivisibles really were available to us. They make dealing with the messy uncertainty of handling curved shapes so much easier. Take the problem of calculating the area of a circle. If there were indivisibles, we could work out the area of a circle very easily. Imagine dividing the circle up into a series of segments, like a slice through an orange. Each is almost a nice, regular triangle, except the short line at the base of the triangle has a slight curve to fit it to the circumference of the circle.

Let's imagine separating each of these segments out of the circle and piling them up, alternating directions, so the curved end of each wedge is on the opposite side to the one below it.

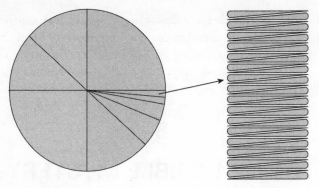

Figure 8.1 From circle to (almost) rectangle.

Then, if you could ignore the little wiggles at the edges, you've built a rectangle.

It's as wide as the radius of the circle – let's be conventional and call it r – and it's as high as half the circumference of the circle (because half the little wiggles are up one side, and half up the other) – so it's $\frac{1}{2} \times 2\pi r$ in height. Or rather, it would be that high if the little wiggles were straight. So we've a rectangle that's πr high and r wide – the area is the height times the width, $\pi r \times r$, or πr^2. Ring any bells?

This proof occurred to Nicholas of Cusa, a fifteenth-century cardinal and philosopher, inspired by a rather similar way that Archimedes had proposed chopping up a sphere into thin slivers to find its volume. It was for his success in supporting the jubilee declared by Pope Nicholas V in 1450, that Nicholas is primarily remembered by the Church. He was sent as a legate to Germany and England, and though in the end he never got across the English Channel, in those parts he reached, Nicholas proved a very effective representative, bringing some of the outlying parts of the Church back into line with Rome, weakening the early beginnings of the Protestant movement and converting many with his powerful sermons.

However, this peaceful man, born in 1401 in the German town

of Cues on the Moselle river near Trier, was certainly much more than an evangelical churchman. His philosophical interests encompassed mathematics, and in his forties he also developed a fascination with astronomy buying sixteen books on the subject (an impressive library in those days) along with astronomical tools. Based on the observations he made and his studies of ancient texts he came up with the most remarkable set of astronomical predictions, ideas that mostly would not be given credence for hundreds of years.

Well before Copernicus and Galileo, Nicholas of Cusa argued that the picture of the Sun orbiting the Earth was topsy-turvy, and it was in fact the Earth that travelled around the Sun. This was, as we have seen, a theory that went back to the Greek philosopher Aristarchus, but Nicholas went further. He also suggested that the stars were not the much smaller lights everyone took them to be, but actually other suns, far distant. And as we have only recently been able to prove, he thought that some of those distant suns would have their own planets circling around them. (Admittedly he also predicted that some of these would be inhabited, something we have yet to establish. In a way, Nicholas's ideas not only prefigured modern astronomy, but science fiction too.)

When it came to the mathematics of the infinite, Nicholas was not foolish enough to suggest that the segments of a circle would ever quite become triangles. Of course, it doesn't work, because those orange segments aren't quite straight. If they were, we could never put them back together to form a circle. In fact Nicholas even used them to illustrate that no matter how much detail we go into, we can never reach the absolute truth of God. But if you *could* take enough of them, an infinite number of them, then it seemed reasonable that the distinction between segments and triangles would be bridged. And after all, the answer comes up correct every time, a fact that is enough for the pragmatist, even if it causes the theoretician to wince.

In fact neither Nicholas nor even Archimedes before him had

been the first to suggest using this type of approach. An ancient Greek philosopher Antiphon, who was a contemporary of Socrates (Socrates was born around 470 BC, around 200 years before Archimedes) said that by drawing any regular polygon (a square, for example) in a circle, then drawing an octagon inside the circle by making equal-sided triangles in each of the four segments of the circle, and so on,

> until the whole area of the circle was by this means exhausted, a polygon would thus be inscribed whose sides, in consequence of their smallness, would coincide with the circumference of the circle.[58]

This view was quickly countered by others who argued that the polygon could never exactly coincide with the circle even if it were possible to carry on the division of the area 'to infinity' (whatever they meant by that, bearing in mind the Greek antagonism to *apeiron*) – and this was the view that would largely be supported by mathematicians ever since.

In the everyday world we take the pragmatist's view every time. We use whatever gets the job done. It doesn't matter that we don't understand how a car or a computer works. We don't need proof. As long as it behaves as expected, we'll use it. In mathematics things aren't that simple. It's easy enough to make assumptions based on the way that things tend to work, and then come a cropper when you finally reach an exception. Where in the sciences we are normally prepared to take a series of measurements and make a deduction, mathematicians are aware that you can't always assume things continue the same way for ever.

Mathematicians need to be supernaturally precise. And so it is inevitable that they would ask just what these indivisibles were. A common view is that the historical mathematicians who worked with indivisibles were being unusually vague – that they weren't clear about what their 'indivisibles' represented. In his excellent book on the history of the number *e* (another transcendental number like π that features frequently in nature, but has no 'exact' value as a ratio of two other numbers, or as a finite equation of squares or other powers), Eli Maor writes:

> Whereas Archimedes was careful to use only finite processes – he never explicitly used the notion of infinity – his modern [i.e., seventeenth-century] followers did not let such pedantic subtleties stand in their way. They took the idea of infinity in a casual, almost brazen manner, using it to their advantage whenever possible. The result was a crude contraption that had none of the rigor of the Greek method that somehow seemed to work: the method of indivisibles ... the method was flawed in several respects: To begin with, no one understood exactly what these 'indivisibles' were, let alone how to operate with them. An indivisible was thought to be an infinitely small quantity – indeed a quantity of magnitude 0 – and surely if we add up any number of these quantities, the result should still be 0...[59]

We will return to whether or not this is true, but first, let's bring in a few more uses of the indivisible. Although Nicholas of Cusa (and Archimedes) had made use of a form of indivisibles, it was in the seventeenth century that indivisibles really took off. In 1615, Johannes Kepler, who became better known for his laws of planetary motion, wrote a work called *Nova stereometria doliorum vinariorum* (A new solid geometry of wine barrels – Kepler didn't confine his interests to astronomical matters), in which he took up Nicholas of Cusa's technique of dividing an area or volume into an infinite number of parts.

Kepler's work inspired the mathematician Bonaventura Cavalieri to develop his magnificently titled *Geometria indivisibilibus continuorum nova quadam ratione promota* (A certain method for the development of a new geometry of continuous indivisibles – note those indivisibles creeping in). In this, he imagined a line (which is one dimensional) to be made up of an infinite number of points, while a flat two-dimensional surface, a plane, was made up of an infinite number of lines, and so on.

Other mathematicians – Pascal and Fermat among them – refined Cavalieri's ideas, moving away from the uncomfortable concept of a collection of lines (say) with no width at all. As Maor said, you can add as many zeros as you like and still get zero. Instead the newer proponents of indivisibles made up a plane by dividing it into an infinite number of infinitesimally wide rectangles, shapes that had some incredibly small width rather than a zero width. The distinction is subtle, but shifted the concept away from the impossible.

It was John Wallis, the inventor of the symbol ∞, who formalized the theory of indivisibles and explicitly stated why it was necessary to move away from Cavilieri's idea of dividing a line into points, a plane into lines, and so on. Wallis dealt with a line (or rather an extremely thin rectangle) that was 'dilutable', having just enough thickness so that by collecting an infinite number of them together it would make up the required size of plane. This insight (and perhaps also the fluid feel of the term 'dilutable') would lead directly to Newton's development of calculus, which is covered in the next chapter.

The indivisible would also feature in the writing of Gottfried Wilhelm Leibniz, Newton's rival in the invention of calculus. Leibniz's indivisible was very much a continuation of the work of Wallis. Could a brilliant mathematician like Leibniz, as Maor suggests, work with a quantity that he did not understand in such a cavalier way? Recent research by Professor Eberhard Knobloch of the Institut für Philosophie in Berlin suggests otherwise. In a paper for the journal *Archive for the History of Exact Sciences*, Professor Knobloch compares the approaches taken by both Galileo and Leibniz on indivisibles and concludes that, though their ideas differed, they each had a clear idea of what they meant – and in neither case was it a matter of adding together an infinite number of zeros.

When Galileo handled the problem of the two concentric wheels, covering the same path length despite their different circumferences, he moved from a many-sided wheel with a finite number of divisible sides, to the circle with its infinite non-divisible sides. The distinction between the two types of sides, defining indivisibles from Galileo's viewpoint was the fact that the circle's sides were non-measurable – that they had no defined size. It wasn't zero, but neither was it measurable. The many-sided wheel he described as *finite lati quanti e divisibili* (finite sides, quantifiable and divisible), while the circular wheel had *infiniti lati non quanti e indivisibili* (infinite sides, not quantifiable and indivisible).

These '*non quanti*' sides were not *infinitely* small, but immeasurably small. We have a tendency to use terms loosely – when we say 'immeasurably small', we normally just mean extremely small. To Galileo, though, this was something so small that it was not capable of being measured, no matter how sophisticated your instrument.

What's more, Galileo's indivisibles were not capable of being combined using normal arithmetic, as arithmetic depends on having some known quantity to deal with – when you lose the concept of quantity you also lose the mechanics of arithmetic. Galileo's description of these indivisibles is quite clear and well thought through. It is entirely possible, as Knobloch suggests, that a lot of the feeling that exists, that those who were using indivisibles did not have a clear picture of what they were dealing with, comes from the fact that the translations of Galileo's work have struggled with providing an adequate rendering of his carefully chosen words. For example, these *non quanti* have often been rendered as 'infinitely small', which entirely misses the point.

By the time Leibniz came to consider indivisibles, much mathematical water had flowed under the bridge. He was to cover them at length in his treatise *De quadratura arithmetica circuli ellipseos et hyperbolae cujus corollarium est trigonometria sine tabulis* (On the arithmetical quadrature of circles, ellipses and hyperbolae), which was written towards the end of the period when Leibniz was working in Paris, around 1676. (Quadrature was the process of calculating the area under a curve, what we would now call integration, or of producing a square of the same area as another shape.) Where Galileo's understanding of indivisibles seems to have been hidden by poor translations, Leibniz has been misinterpreted thanks to a lack of publication. Remarkably, in what must be something of a record, this treatise, the longest Leibniz ever wrote, was not published until 1993, some 317 years after it was written.

This was the work of one of the most remarkable mathematicians of the seventeenth century. Born in Leipzig in July 1646, Leibniz came from an academic family, his father a well-established lecturer in moral philosophy in his home city. Like his near contemporary Isaac Newton in England, Leibniz found himself being educated in a system that still placed much value on the systems bequeathed by the ancient Greeks, particularly Aristotle. Also like Newton, Leibniz began to question the findings of the ancients from an early age, wanting to work things out himself, rather than depend on received wisdom.

Initially it was philosophy that was Leibniz's primary interest – mathematics was not well taught at his first university, Leipzig. After his initial degree he moved around frequently, living briefly in Jena, Altdorf, Franfurt and Mainz, expanding his interests beyond his nominal subjects of philosophy and law to take in science and mathematics. It was at Jena, particularly, that Leibniz first became fascinated by mathematics, influenced by his professor, Ehard Weigel. But in 1672, still only in his twenties, he was sent to Paris on a diplomatic mission on behalf of his sponsor Baron Johann Christian von Boineburg. There he was to stay only four years. While his role was technically diplomatic, he vastly increased his range of scientific and mathematical contacts both in France and England (he was made a Fellow of the Royal Society in 1673) and laid the foundations for much of his best work.

Leibniz had attempted to publish his paper on indivisibles while still in Paris, but xenophobia got in his way. He was unable to continue at the Academy of Sciences as it was considered that there were already enough foreigners there. He was, however, able to obtain the post of librarian and Court Councillor to the

Duke of Hanover, and returned to Germany. At the time it may have seemed like just another step in his nomadic life, building experience as he travelled through Europe, but Leibniz was to remain in Hanover until his death in 1716. Before he left Paris, Leibniz deposited the fair copy of his paper with a friend, but the friend died before he could do anything with it. Eventually the paper was sent on to Leibniz in Hanover, only to get lost in transit.

At this point, Leibniz could have reworked his original, scrawled manuscript, but by now it seemed more than a little dated. He had written it before he developed the notation that he would use so successfully in his development of calculus (still used today), and probably felt that despite its useful insights, it was not worth the effort to bring it up to date. The paper languished for many years, in part because Leibniz's handwriting was difficult to decipher (this was only ever intended as a rough, personal copy), until it was finally edited by Professor Knobloch and published in 1993.

In his treatise, Leibniz uses the method of indivisibles to find the areas of spaces by taking the 'sums of lines'. These weren't truly lines, because the method entailed (as Wallis had made clear) adding together rectangles with 'equal breadths of indefinite smallness'. Leibniz shows that it is possible to construct shapes from a series of these slices of varying height, each differing from each other (or the shape being constructed) by an amount that is smaller than any given quantity. He points out that you can get within any desired limit of matching a shape this way, even when the number of rectangles is finite.

Leibniz's careful detail here is entirely contrary to Maor's suggestion that 'no one understood exactly what these 'indivisibles' were, let alone how to operate with them'. It is true, though, that despite being a superb mathematician, Leibniz himself found the detail he had to work to quite irritating. He commented that he did not want the 'excessive exactness' to discourage the reader's mind from other far more agreeable things by making it weary prematurely. He went on to comment:

> Nothing is more alien to my mind than the scrupulous attention to minor details of some authors which imply more ostentation than utility. For they consume time, so to speak, on certain ceremonies, include more trouble than ingenuity and envelop in blind night the origin of inventions which is, as it seems to me, mostly more prominent than the inventions themselves.[60]

The indivisibles that Leibniz worked with, unlike Galileo's *non quanta*, did have a measurable magnitude. They were infinitely small quantities, but quantities nonetheless – quantities that were defined by being smaller than any given quantity you would care to specify. In a sense these quantities were fictional, not having any true parallel in reality (because they were smaller than any 'real' quantity), yet Leibniz, building on the work of his predecessors, had moved from Galileo's incalculable non-quantities to something that could be handled with mathematics, that could be made part of a calculation. Leibniz would later say, in answer to concerns about using the infinite, that it was not necessary to deal with the infinite in strict terms, but it was more in the nature of an analogy, he was dealing with an unreal quantity to produce a real result.

Even so, the image of indivisibles has largely remained as one of an imprecise, pragmatic use of a woolly idea that works. Working with infinity would always be a risky business. Leibniz, pointing out the ease with which the unwary can slip into absurd results when dealing with infinity says that 'calculation with the infinite is slippery'. A biographer of John Wallis later commented on the practice of dividing and multiplying by infinity:

> For many years to come the greatest confusion regarding these terms persisted, and even in the next century they continued to be used in what appears to us as an amazingly reckless fashion.[61]

Though it has proved to be the case that Galileo and Leibniz had a much better idea of what they meant than we have given them credit for, it is certainly true that some manipulators of indivisibles were less than formal in their approach. But indivisibles were not to remain an entertaining intellectual challenge. A new formalization of the use of quantities that were so small that they could almost be considered non-existent was to cause the outbreak of a bitter, three-way intellectual battle. Two of the contenders in this fight were Leibniz himself and Isaac Newton. The third, unlikely contestant was an Irish Anglican bishop.

9

FLUXION WARS

Suffering is permanent, obscure and dark,
And shares the nature of infinity.

William Wordsworth, *The Borderers*, III.1539

FOR SOMETHING THAT WAS REGARDED AS NON-EXISTENT, infinity was capable of producing very real emotions. In the seventeenth century it precipitated a battle between three individuals that would today have resulted in spectacular lawsuits – and though there was no legal involvement in this case, the dispute was no less acrimonious. The first contender in the battle of the infinite was Sir Isaac Newton.

Newton's first thoughts on the matter took place not in the refined quiet of his Cambridge college, nor in the halls of the Royal Society. Instead it was in a rural farmyard at Woolsthorpe in Lincolnshire. Young Isaac was born into a family that teetered on the borderline between gentility and rustic farming. Although his family was never poor, his early years were marred by the death of his father some months before Isaac's birth on Christmas morning 1642. Once Newton's mother remarried to a local rector she had little interest in keeping up a direct involvement in the day-to-day business. It was not entirely surprising that she tried to shield Newton from the academic world that was soon attracting him, hoping to keep him at Woolsthorpe to manage the farm.

Hannah Newton was not to succeed in keeping her son a country boy. His teachers and other relatives all put pressure on his mother to allow him to continue his education after academic success at the nearby King's School in Grantham. Hannah grudgingly allowed Newton to go off to Cambridge (though she refused to give him more than a minimum of financial support). Yet in 1664, despite his clear reluctance to be involved in affairs at Woolsthorpe, he returned home for two years. This effective exile on the farm was brought about not by Hannah's persuasion, but by the plague.

By then, Newton was 23. He had graduated with an acceptable but not outstanding Batchelor of Arts degree and was working towards his MA. He had no desire to return home, but the bubonic plague that was already starting to devastate London spread to Cambridge, and with the other members of the university, Newton was asked to leave until the outbreak had died down. The next two years in rural isolation would prove to be one of the most productive times of Newton's life. He performed experiments on the nature of light and colour. He explored in his mind the way that gravity took its effect and attempted to describe rules that would explain the motion of the planets around the Sun.

Newton's ideas on gravity and planetary motion were developing well, but he lacked the tools to be able to work properly on the numbers involved. It was on the farm at Woolsthorpe, according to the well-developed legend which sprang up soon after his death, that Newton, sitting idly in the orchard, watched an apple fall to the ground. (The business of its falling on his head has always been recognized as apocryphal.) Although the event itself is legendary, it is very likely that at some point Newton did see an apple fall – whether it spurred him into thought at the moment, or simply remained as a subconscious irritation like the grain of sand that starts a pearl in an oyster is neither certain nor really important.

The fact is that Newton thought about the workings of gravity.

As an apple dropped from the tree, how did its speed vary as it fell towards the ground? How could the movement of the planets in their orbits be predicted? At the heart of these problems was the need to be able to predict the rate at which something was undergoing change.

Imagine a rocket, blasting off into space. To begin with it hardly seems to move it all: it claws its way ponderously off the launch pad. Soon, though, it has accelerated. It is moving faster and faster through the sky. You could imagine drawing a graph that showed just how fast the rocket was going, measuring its speed at each moment. Initially it might look something like Figure 9.1.

To make this example easy I have used a very special graph, with a fixed relationship between the time that has passed by and the speed at which the rocket travels. At any particular time, to find out the speed of the rocket you simply have to take the square of the time that has passed. Not only is the rocket getting faster and faster, but the rate at which it is getting faster is growing too. Unlike Newton's apple, which was under the constant pull of the Earth's gravity, our rocket has an increasing rate of acceleration. (I'm not saying a real rocket necessarily behaves like this, I am just using these figures to make the example obvious.)

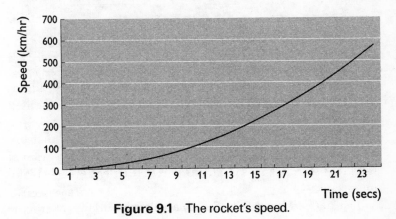

Figure 9.1 The rocket's speed.

So, as far as our rocket is concerned, at any point in time we can easily work out the speed because

$$\textbf{speed} = \textbf{time} \times \textbf{time},$$

or **speed** = **time**^{2}.

Ten seconds into the journey, the speed is 10 × 10, or 100 kilometres per hour.

This is fine, but Newton wanted to know more. Not just how fast something was going, but what the acceleration was. What was the rate at which the speed was changing? This would be incredibly easy if we were dealing with a straight line instead of a curve.

If we had a nice, steadily accelerating rocket, going at the speed shown in Figure 9.2, then we could work out the acceleration easily, because the acceleration is exactly the same as the slope of the line. When we look at a hill, the slope tells us how fast the hill is climbing, just as the acceleration tells us how fast the speed is climbing.

If we look at the slope of a 1 in 4 hill – it rises up by 1 metre for every 4 metres you travel forwards – we can say that the slope is 1/4 or 25%, as you'll now see on road signs. The slope is the change in distance upwards, divided by the change in distance

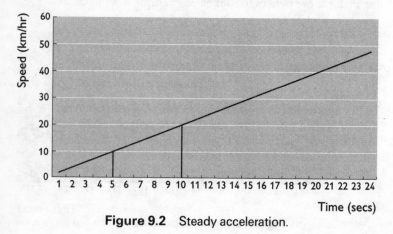

Figure 9.2 Steady acceleration.

forwards – that's simply what we mean by 'slope'. For example, in Figure 9.2, in the time between 5 and 10 seconds having elapsed, the rocket has gone from travelling at 10 kilometres per hour to 20 kilometres per hour. It accelerated by 10 kph in 5 seconds, or 2 kph per second. The rate at which it gets faster is exactly the same as the slope of the graph, which is rising at 2 units for every 1 it travels along from left to right.

With a curve, we can't just work out the slope by taking the difference between two points, as we did with the straight line, because the slope of the curve is always changing. But what we can do is to work out the slope at a particular point by drawing a line across the 'outside' of the curve that fits symmetrically, snugly against the curve. This line, called the tangent, gives an instantaneous view of the slope – of the acceleration.

In devising calculus, Newton was providing a simple way to work out just what that slope – and hence the rate of change – would be.

Newton's argument made a clever use of our slippery little infinitesimals. Let's go back to the first curve we used to show the rocket's acceleration. This graph, as we've seen, can be completely described by a short equation: **speed** = **time**^{2}.

Newton wanted to know what the acceleration was at any particular time. What was that elusive tangent for any point on the curve? To get an answer, Newton took a quantity that was very like the indivisibles discussed in the previous chapter. Unlike Galileo's *non quanta*, Newton's extremely small quantity was capable of being used in calculation. He called it o – a sort of misshapen zero that's not quite zero but just a little bit more. If we start from anywhere we like on the rocket's speed graph and move on a tiny amount of time, we could say that we've moved from **time** to **time** + o. Now, **speed** (which remember is **time**^{2}) becomes (**time** + o) × (**time** + o) – so how do we get to that rate of change we're interested in? Remember it's the same as the slope of the tangent.

Now, Newton argued, in such a tiny change the curve can be

considered to be straight, a microscopic tangent in its own right. This means that we can work out the slope just as we did before from the change in distance upwards divided by the change in distance forwards. In this case, that means that the slope that defines acceleration is the change in speed ('up' on the graph) divided by the change in time ('forwards' on the graph). We have said that the speed goes from $\textbf{time}^2$ to $(\textbf{time} + o) \times (\textbf{time} + o)$, or simply $(\textbf{time} + o)^2$, so the change in speed is

$$(\textbf{time} + o)^2 - \textbf{time}^2,$$

while the change in time is $(\textbf{time} + o) - \textbf{time}$.

So the slope, or the acceleration, is

$$\frac{(\textbf{time} + o)^2 - \textbf{time}^2}{(\textbf{time} + o) - \textbf{time}}.$$

Let's multiply everything out and get rid of those brackets. We get

$$\frac{\textbf{time}^2 + \textbf{time} \times o + o \times \textbf{time} + o^2 - \textbf{time}^2}{\textbf{time} + o - \textbf{time}}$$

or, even simpler,

$$\frac{2 \times \textbf{time} \times o + o^2}{o}.$$

Dividing out those o's gives

$$2 \times \textbf{time} + o.$$

If we imagine o getting smaller and smaller, as it becomes nearly 0, then to all intents and purposes this becomes $2 \times \textbf{time}$.

Newton termed his calculus 'the method of fluxions', where his 'fluxion' was the rate of change of something (the original value that was changing he called a 'fluent'). There's a very important reason for these terms that sound more like the products of a quack medicine vendor than mathematical expressions. Newton was aware, as his critics would soon vociferously point out, that there was a real danger with his disappearing o. Remember that the equation involved dividing the infinitesimal quantity o by itself. If this ever becomes zero divided by zero, the whole thing falls apart.

The business of dividing zero by zero has caused much confusion over the years. Any other number divided by itself comes up with one. Yet zero divided by any other number is still zero. And divide any normal number by something that is very close to zero and the result gets closer and closer to infinity. To see the confusion caused by this unholy ratio, you only have to look at the attempts at explaining it by two early Indian mathematicians. One, Brahmagupta, writing in the seventh century, decided that zero divided by zero was zero, while in the twelfth century, Bhāskara thought that anything divided by zero, including the result of 0/0, was infinite. In reality, 0/0 is indefinable. It is anything and nothing. It has no useful value.

Newton's escape from the perils of this division was to hide in his terminology. Instead of thinking in terms of very small (increasingly small) parts, he thought in terms of motion, of fluidity. Remember how John Wallis had written of his indefinitely thin rectangles as being dilutable. For Newton, his value o was in the process of being diluted. It wasn't so much a value as something that was in the process of flowing away to nothing, like the last drops of water disappearing down a plughole. When he was looking at a curve, he thought of this as the continuous motion of a point and so there was no need to worry about the reality of division by zero, even though the mathematical symbols used seemed to imply it. What was important in terms of calculus, Newton suggested, was not what *was*, but how it moved and changed – not the infinitesimal but the minuscule motion. (It's entertaining to think that this is the absolute reverse of the view of Zeno and his colleagues, who believed moving and changing to be an illusion, and it was only what *was* that mattered.)

Newton's invention of calculus (we have only covered a part of it so far, but we've seen enough to get a feel for the way the infinitely small comes into it) was to revolutionize mathematics, science and engineering. It was to become a mainstay of technology. So how is it that that this word 'fluxion' is to all intents and purposes no longer part of the language? My dictionary calls it 'rare or obsolete'. Most of us who have taken calculus even to the levels used in university science and engineering will never come across the term. Enter the second party to the fluxion wars – our old friend, Gottfried Wilhelm Leibniz.

Leibniz too devised a form of calculus. The effective workings of Leibniz's calculus were very similar to those of Newton's, although he did take a more straightforward view on his indivisible infinitesimal quantities, without the need for Newton's insistence on the fluidity of his values. Leibniz's wording was quite different, but the fundamentals of the two founders of calculus were very similar. It might seem that this was quite reasonable. They both followed the same tradition linked back through Nicholas of Cusa to Aristotle. Yet it's not entirely surprising that Newton suspected Leibniz of stealing his ideas. And he was eventually to make this a very public claim.

As we have seen, Leibniz came over to London in 1673 while he was living in Paris, and become closely involved with the Royal Society, a link that would remain active for the rest of his life. He was in regular correspondence with the big names of the society, including the secretary, Henry Oldenburg, and the curator of experiments, Robert Hooke. Newton had been elected a Fellow in 1672, and was also in regular correspondence with Oldenburg. Hooke, though, proved anything but a friend to the young Cambridge scientist. Not only did he dispute Newton's findings and claim to have invented the reflecting telescope before him, but he used his place in the social in-crowd to emphasize Newton's lack of social graces. For a while, Newton was disillusioned with the Royal Society as a whole.

Now when Leibniz was putting together his eventually abandoned treatise, he had written to Oldenburg to tell him that he was thinking of such work. Oldenburg, not aware that Leibniz had more than a passing interest, informed him of what he knew of Newton's thinking at the time. Oldenburg also put Leibniz in touch with the publisher and Fellow of the Royal Society John Collins, a man who had a continued interest in Newton's work and to whom Leibniz now wrote on a number of occasions as a sounding board for his theories. Whether Newton suspected Oldenburg, Collins or the infuriating Hooke of giving away information to his rival, he wrote to Leibniz via Oldenburg, at Oldenburg's suggestion, in June 1676, describing the results he had so far achieved without giving him any useful reasoning.

There was, in fact, a condensed explanation of the method of fluxions in the letter, but Newton, wary of his ideas being stolen, wrote:

> I cannot proceed with the explanation of the fluxions now, I have preferred to conceal it thus: 6accdæ13eff7i3l9n4o4qrr4s8t12vx.[62]

This apparent gibberish, more like the code we now type to unlock a piece of software than useful information, defines the relationship of the fluent and the fluxion – hardly definitive in proving precedence, but giving Newton some sense of security. He took the phrase *Data æquatione quotcunque fluentes quantitates involvente, fluxiones invenire: et vice versa* (Given an equation that consists of any number of flowing quantities, to find the fluxions: and vice versa) and counted the frequency with which each letter occurs, then strung them out in alphabetical order. If there were three or more, instead of listing them he put a number in front (so there are 9 n's and 4 q's). The apparently excessive number of v's is accounted for by the lack of distinction in written Latin between U and V.

Unfortunately, Newton's suspicions that Leibniz was stealing his ideas on the method of fluxions were to be fuelled by the slowness of the carriage of letters at the time. Although Leibniz responded as soon as he received Newton's letter, describing his own thoughts on the calculus, weeks appeared to have passed by before he bothered to respond. Weeks in which he could have taken Newton's ideas and uncovered some of the working behind them – or at least this is how Newton seems to have interpreted it.

Newton's second letter, more guarded than the first, was limited by the formal politeness of the day that made it necessary to phrase the greatest of insults in a roundabout manner. Take, for example, a letter Hooke wrote to Newton in which he claimed Newton had stolen his ideas – the strength of feeling hardly comes across in Hooke's words of accusation:

> I . . . am extremely well pleased to see those notions promoted and improved which I long since began, but had not time to complete.[63]

Newton now asserted much the same of Leibniz. If his anger was fuelled by delays over the weeks the previous letter had taken to travel, it must have brought him to bursting point to wait for a reply this time, as the letter took eight months to catch up with Leibniz who had moved to Hanover while it was en route. Leibniz did reply, trying to demonstrate how his thinking was original, though Newton dismissed what he had written as not showing anything that was not already known.

This feeling of being robbed by Leibniz would remain with Newton for the rest of his long life (he was 34 when the correspondence with Leibniz began, and died at the age of 85). This near-paranoia was not helped by the fact that Leibniz's enhanced description of calculus was first formally published in 1684, well ahead of Newton's fluxion-based theories, which didn't reach print in English translation for nearly 60 years (Newton wrote the paper in Latin in 1671, and first published it in Latin along with his *Opticks* in 1704).

For a while it seemed as if Leibniz would be above the dispute and would simply ignore Newton's insistence that the German mathematician was guilty of stealing ideas, but Leibniz felt that he had to come to his own defence when others began to slant the story in Newton's favour. Accusations flew backwards and forwards across the channel, strengthened by Leibniz's publication. Finally, in 1708, the Scottish mathematician John Keill made the dispute a totally public one. He wrote a paper for the *Philosophical Transactions* of the Royal Society that explicitly accused Leibniz of plagiarism. This was already more than 30 years after Newton's original letters. Yet when challenged by Leibniz, Keill pointed back to the letters from Newton, suggesting that Leibniz had – or, at least, could have – derived the principles he put forward as his calculus (a term that Leibniz derived from the Latin name for the small stones used in a simple abacus).

Leibniz now complained to the Royal Society – he was still a Fellow, after all. Eventually, in 1711, his letter was read out, and Keill was asked to write an apology. But there is no doubt that Keill was very strongly under Newton's influence, and it has even been suggested that Newton was ghost writer for a reply in which Keill remarked that he had only suggested that

> Mr Newton was the first discoverer of the arithmetic of fluxions or differential calculus.

He goes on to point out that Leibniz could have derived his principles from Newton's letters and then gives the weak sop that Leibniz ('that illustrious man')

> did not need for his reasoning the form of speaking and notation which Newton used, he imposed his own.[64]

Leibniz could hardly take this letter which only admitted the originality of his notation and terminology as an apology, and came back to say that it was unacceptable that his sincerity should be attacked in this way. After all,

> no fair-minded or sensible person will think it right that I, at my age and with such a full testimony of life, should state an apologetic case for it, appearing like a suitor before a court of law.[65]

After writing this letter, the 66-year-old Leibniz probably felt that at last he had got his point across. Soon after it was announced that the Royal Society would set up an inquiry, a committee of eleven men, to establish the truth. However, this examination of the rights and wrongs of the Great Priority Dispute (it became famous enough to warrant capital letters) was to be no work of blind justice, impartially weighing the scales of probability. The outcome of the inquiry, written by no less an eminent figure than the Royal Society's president, was firmly biased against Leibniz. In fact, the committee did not even bother to take any evidence from the German. The report's author? Isaac Newton.

Leibniz could not beat the system, at least not with such a masterful manipulator as Newton at the helm, and though he continued to protest his innocence both verbally and in writing, he was never formally vindicated. The rift not only caused upset between the individuals involved, but left a divide between the mathematicians of Britain and continental Europe that would take a century to close. The truth of the matter? While it is always hard to be certain looking back from such a distance, it would seem that Newton had arrived at the original concepts sooner, but that Leibniz was not guilty of plagiarism, and developed in parallel a much more practical form than Newton's had been.

Because of this practicality, though there was little to choose between Newton's and Leibniz's approach in terms of results, it was in fact the Leibniz terminology that won the day and is now universally used. We don't speak of fluxions any more (it's just as well – they do sound unfortunately like something you use to clean drains), we learn calculus. Instead of Newton's squashed zero, Leibniz used a variation on an already accepted theme, a variation that has survived unchanged to the present day.

The Greek capital letter delta (Δ) had already been used for a while as a shorthand term to indicate change. The small delta δ was used with what seems unusually approachable logic in the establishment of mathematical symbols to mean a small change. So if you wanted to show that something called **time** had undergone a small change you would call that change δ**time**. Leibniz took this one stage further for calculus where the change was not just small but infinitesimally small, and replaced the delta with a conventional 'd' – so the small change δ**time** became the infinit-esimal change d**time** (most often a single letter like dt or dx), This dt or dx was Leibniz's equivalent of Newton's o.

It's not really surprising that the notation Leibniz developed here became the standard, and the choice certainly doesn't reflect any anti-Newton bias. Using o inevitably caused confusion with zero, and did not indicate what it represents a small increment of. Leibniz's dx (or whatever) was much clearer. Similarly, when writing the rate of change of x, Newton stuck a dot on top of it: $\dot{x}$. This is often called the dot notation, though Newton actually called it 'pricked' in the same sense that written music was often called 'pricked music' to indicate the use of dots or prick marks to show the position of the notes on the stave.

Newton's approach was a nightmare for printers, easily mis-taken in handwriting for a slip of the pen. Leibniz simply made the rate of change of x dx/dt – the change in x divided by the change in time – making the involvement of time explicit. He could then extend this to dx/dy showing how x changed with respect to y, whatever that was (Newton's equivalent would be $\dot{x}/\dot{y}$) – Leibniz's version was much easier to handle.

As well as the aspect of calculus that allowed calculation of tangents and accelerations (so-called differential calculus, as it deals with change, with differences) both men also covered the direct descendent of those collections of 'squashed rectangles' that date back through John Wallis and Nicholas of Cusa to the an-cient Greeks. This 'integral calculus' is typically used to find the area underneath a curve or the volume of a three-dimensional object, and once again it is Leibniz's symbol for this special kind of addition of many slivers that has stayed with us.

Newton had seen integration as a reversal of differentiation, and so had not come up with any special terminology for it. Leibniz explicitly thought of it as a sum, and unlike his simple logic in the transition from δ to d, here Leibniz made the rather less obvious move from an S for sum (or rather *summa*) to a strange, stretched sign that is halfway between an S and a bracket, the integral sign $\int$ – yet despite its oddness, this is the way we still represent these calculations.

Whichever flavour you favoured, calculus worked like a dream. It made what had been impossibly complex calculations trivial. In effect, in the late seventeenth century, Europe underwent a 'calculus revolution' that had an impact on science and engineering similar to that of the later IT revolution, where once again a new technical capability would make what had been impossibly complex into something remarkably simple. But there was a snag.

Just as was the case with Nicholas de Cusa's approximation dividing a circle into orange segments which were then approximated to triangles, there seemed to be a fudge in calculus. It's easy to feel uncomfortable about the way this o (or dt) is used. After all, to get to our $2 \times$ **time** $+ o$ (or $2t + dt$) above we had to divide by o (or dt). But if you only get the exact result when o becomes 0, you're dividing by 0 and all bets are off.

It was this that really worried Bishop Berkeley, the third and final entrant in the fluxion wars. Berkeley was neither a supporter of Newton nor of Leibniz – in fact he was ferociously opposed to both.

If Bishop George Berkeley is remembered for anything other than his opposition to calculus, it was his dealing with the problem of whether or not matter or an event truly existed without an observer. His classic example was whether a tree, hidden away in the middle of a forest that no one could possibly be thinking of really could be said to exist (at that time). Berkeley believed that the observer was required, though he then effectively negated the value of the theory by pointing out that there always *was* someone present, God, so yes the tree did always exist. Even so, Berkeley was an insightful philosopher whose objections to the work of Newton and Leibniz might have originally arisen from religious convictions, but were nonetheless based on some sensible mathematical reasoning.

It's a natural reaction to think that a bishop objecting to a scientific theory was some sort of old fogey, behind the times and not appreciating the distinctive ideas of the new generation of scientific thinkers. In fact Berkeley, born at Dysart Castle, outside the small town of Thomastown in Ireland on 12 March 1685, was around 40 years younger than Leibniz and even more Newton's junior (though admittedly Berkeley was already 49 when his critique of calculus was published).

The ruins of Dysart Castle still stand in County Kilkenny, though not the outbuildings in which Berkeley's family lived. It is tucked away in the centre of the county, very much a back-water. But Berkeley's family were no country bumpkins. They were recent incomers, leaving England under a cloud after the Restoration. At the time, with this background, it would have been unlikely for Berkeley's life to be limited to this lovely setting, or to Ireland alone – and such proved to be the case. He was educated at Trinity College, Dublin, and became a Fellow in 1707. The next six years were to be the longest continuous period he would spend in Ireland after graduation. He was ordained into the Anglican Church, and moved to London in 1713.

This, though was only the beginning of the future bishop's travels. He took a great interest in the development of the colonies, and spent some time in working up a plan for a college in Bermuda. In order to try to put his plans into practice, he sailed over to the Americas, buying a house in Newport, Rhode Island, which he used as a base for the three years in which he worked on funding for the project. But there was not enough money available, and Berkeley moved back to London. It seems that, despite his failure to win support financially, he was in good social standing. According to a biography written by Joseph Stock in 1776:

> After Dean Berkeley's return from Rhode Island, the Queen often commanded his attendance to discourse with him on what he had observed worthy of notice in America. His agreeable and instructive conversation engaged that discerning Princess so much in his favour, that the rich Deanery of Down in Ireland falling vacant,

he was at her desire named to it, and the King's letter actually came over for his appointment. But his friend Lord Burlington having neglected to notify the royal intentions in proper time to the Duke of Dorset, then Lord Lieutenant of Ireland, his Excellency was so offended at this disposal of the richest Deanery in Ireland without his concurrence, that it was thought proper not to press the matter any further. Her Majesty upon this declared, that since they would not suffer Dr. Berkeley to be a *Dean* in Ireland, he should be a *Bishop*; and accordingly in 1733 the Bishopric of Cloyne becoming vacant, he was by Letters Patent dated March 17, promoted to that see, and was consecrated at St. Paul's church in Dublin on the 19th of May following by Theophilus Archbishop of Cashel, assisted by the Bishops of Raphoe and Killaloe.[66]

It was about the time of his move to the Manse House in Cloyne (where according to his biographer he constantly resided except for one winter that he attended the business of Parliament in Dublin), that Berkeley became embroiled in the calculus debate. The trigger to Berkeley's involvement was the input of Edmond Halley. We tend to think of Halley purely as an astronomer (he was the discoverer, as Astronomer Royal, of the famous Halley's comet), yet he was also Savilian Professor of Geometry at Oxford and personally responsible for pushing his friend Newton into getting his great *Principia* into print – in fact Halley published it at his own expense.

However, whereas Newton was a Christian (admittedly somewhat erratic in his church attendance), Halley was a very vocal agnostic. In Berkeley's contemporary biography we hear how he was said to have taken against Halley. A friend of Berkeley's, one Mr Addison, had been to visit a dying friend, Doctor Garth. When Addison began to speak to the Doctor about making preparation for meeting his maker, Garth replied:

> Surely, Addison, I have good reason not to believe those trifles, since my friend Dr. Halley, who has dealt so much in demonstration, has assured me, that the doctrines of Christianity are incomprehensible, and the religion itself an imposture.[67]

This was not a challenge that Berkeley could treat lightly. The bishop may well have got round to putting his doubts about fluxions into print anyway, but it is thanks to Halley's input that his work on the subject comes across so strongly, and has the unequivocal and rather magnificent title *The Analyst: A Discourse Addressed to an Infidel Mathematician*.

Berkeley begins with a personal assault on that infidel Halley's wisdom in commenting outside his normal field:

> Though I am a stranger to your person, yet I am not, Sir, a stranger to the reputation you have acquired in that branch of learning which hath been your peculiar study; nor to the authority that you therefore assume in things foreign to your profession; nor to the abuse that you, and too many more of the like character, are known to make use of such undue authority, to the misleading of unwary persons in matters of the highest concernment, and whereof your mathematical knowledge can by no means qualify you to be a competent judge.[68]

Berkeley then girds his loins to take on fluxions. He begins by admitting their importance to modern mathematics:

> The method of Fluxions is the general key by help whereof the modern mathematicians unlock the secrets of Geometry, and consequently of Nature. And, as it is that which hath enabled them so remarkably to outgo the ancients in discovering theorems and solving problems, the exercise and application thereof is become the main if not sole employment of all those who in this age pass for profound geometers. But whether this method be clear or obscure, consistent or repugnant, demonstrative or precarious, as I shall inquire with the utmost impartiality, so I submit my inquiry to your own judgment, and that of every candid reader.[69]

Then he defines the fluxions which he is about to attack:

> ... a method hath been found to determine quantities from the velocities of their generating motions. And such velocities are called fluxions: and the quantities generated are called flowing quantities. These fluxions are said to be nearly as the increments of the flowing

quantities, generated in the least equal particles of time; and to be accurately in the first proportion of the nascent, or in the last of the evanescent increments.[70]

In other words, these fluxions aren't really there at all, but are rather the moment of something coming into existence or ceasing to exist. Berkeley then brings in the Leibniz version of calculus, just to make sure that Halley doesn't have an escape route by changing terminology:

The foreign mathematicians are supposed by some, even of our own, to proceed in a manner less accurate, perhaps, and geometrical, yet more intelligible. Instead of flowing quantities and their fluxions, they consider the variable finite quantities as increasing or diminishing by the continual addition or subduction of infinitely small quantities. Instead of the velocities wherewith increments are generated, they consider the increments or decrements themselves, which they call differences, and which are supposed to be infinitely small.[71]

It might seem initially that Berkeley is more in favour of these 'differences' from his 'yet more intelligible' remark, but he is soon putting them in their place:

Now to conceive a quantity infinitely small, that is, infinitely less than any sensible or imaginable quantity, or any the least finite magnitude, is, I confess, above my capacity. But to conceive a part of such infinitely small quantity that shall be still infinitely less than it, and consequently though multiplied infinitely shall never equal the minutest finite quantity, is, I suspect, an infinite difficulty to any man whatsoever; and will be allowed such by those who candidly say what they think; provided they really think and reflect, and do not take things upon trust.[72]

Then Berkeley gets to the heart of his point, one that is often missed in descriptions of his attack on calculus:

All these points, I say, are supposed and believed by certain rigorous exactors of evidence in religion, men who pretend to believe no further than they can see. That men who have been conversant only about clear points should with difficulty admit obscure ones might not seem altogether unaccountable... But with what appearance of reason shall any man presume to say that mysteries may not be objects of faith, at the same time that he himself admits such obscure mysteries to be the object of science?[73]

Berkeley points out that agnostic or atheist mathematicians who say that they can only believe in what they can see – and so have difficulty with religion – are being hypocritical, because they seem quite happy to accept fluxions or differential calculus, which is fundamentally dependent on a concept which it is beyond human imagination to clearly envision. It's a neat argument, with a significant element of truth in it. He then goes on to show how the assumptions underlying calculus are at best shaky, the point being not to undermine calculus per se, but rather to show how unreliable the infidel mathematician's judgements are.

Berkeley gives a range of demonstrations of the inconsistencies and oddities that come out of these infinitesimal quantities in a total of 50 sections to his pamphlet. He works through some typical workings of calculus and then hits out against its problems:

> Let now the increments vanish, and their last proportion will be 1 to nx^{n-1}. But it should seem that this reasoning is not fair or conclusive. For when it is said, let the increments vanish, *i.e.* let the increments be nothing, or let there be no increments, the former supposition that the increments were something, or that there were increments, is destroyed, and yet a consequence of that supposition, *i.e.* an expression got by virtue thereof, is retained... when we suppose the increments to vanish, we must suppose their proportions, their expressions, and everything else derived from the supposition of their existence to vanish with them.[74]

In a nutshell, either the sums didn't work because there was always a tiny fraction of the fluxion left, or o was really 0 and the whole thing didn't make sense. It is part way through a later demonstration that he makes his most poetic and frequently quoted assault on the fluxion:

> And what are these fluxions? The velocities of evanescent incre-
> ments. And what are these same evanescent increments? They
> are neither finite quantities nor quantities infinitely small, nor yet
> nothing. May we not call them the ghosts of departed quantities? [75]

Certainly Berkeley had a point, but it wasn't one that had much
impact on the real world – because calculus did the job. Newton,
Leibniz and the following generations were happy to keep the
infinitesimals in the virtual world of potential infinity, leaving
the philosophical pondering to those who had nothing better to
do.

In fact, Newton was never happy with infinitesimals per se.
Instead he used an argument that would become the principle
tenet of calculus, an argument that made it unnecessary to deal
with those fiddly ghosts. Rather than saying his o was an infinitesimally small number (perhaps $1/\infty$), he would simply have
said that the value of his rate of change in our rocket example
was $2x + o$ and that this 'tended to $2x$ as o tended to 0'. In other
words it got closer and closer to $2x$, though never quite making it,
just like that series we met way back in Chapter 2:

$$1 + \tfrac{1}{2} + \tfrac{1}{4} + \tfrac{1}{8} + \tfrac{1}{16} + \tfrac{1}{32} + \cdots$$

adds up to something that's closer and closer to 2 without ever
quite getting there.

This idea of 'tending to' was given a formal blessing in the mid-
nineteenth century when a pair of mathematicians operating
independently, but with none of the ire of Newton and Leibniz,
took the two steps that were required to move away from the need
to deal with raw infinity. The first contributor was Augustin-
Louis Cauchy. A 32-year-old lecturer at the Collège de France,
Cauchy had already made many mathematical discoveries, but
suffered in his career progression from a lack of political skill. In
1821, in *Cours d'analyse*, a textbook for the students of the École
Polytechnique, he explicitly stated that both infinity and the

infinitesimal were variable quantities – a more formal concept than a fluxion – so that rather than being a true number, infinitesimals were just a label for something that had no fixed value, but got smaller and smaller while never actually reaching zero.

Imagine we were undertaking some form of calculus – it doesn't really matter what. We can imagine it as a black box for turning input into output. That input might be a rate of acceleration that we are turning into speed at a certain time, or it might be the shape of a graph that we are turning into the area under it. The calculus process binds up the input with a very large quantity – the thing that 'tends to infinity' and a very small quantity – the infinitesimal that heads off towards zero.

Using Cauchy's approach we say that instead of infinity we only need use a large enough real number, and instead of 0 we only need a small enough real number. That way, we can make the result very close to the expected output. And for any particular closeness to the desired output we can specify a big enough large quantity and a small enough small quantity. It's as if there was a sliding scale – we just push the marker along towards potential infinity (in the case of the infinitely large) or 0 (in the case of the infinitesimally small) until it satisfies the requirement. Admittedly it isn't entirely clear what Cauchy meant by a variable, but the principle was established.

The final step to pin down the problem of calculus came from Karl Theodore Wilhelm Weierstrass. Like Cauchy, the younger Weierstrass (he was born 25 years after Cauchy in 1815 in Westphalia) had a talent that seemed to lack academic recognition to begin with thanks to his lack of interest in political machinations – and a total absence of enthusiasm for the dull plodding that accompanies the absorption of the basics that students have to learn. Rather than spend much time on his lectures, Weierstrass seems to have concentrated more on drinking with his friends and indulging in his sporting passion, fencing.

Weierstrass failed to get his degree, and instead took a teaching diploma at the Münster Academy, a route into the academic world that was hardly likely to endear him to the authorities. Just as Einstein in the early years of the twentieth century was to do much of his original work while passing the time as a patent clerk, Weierstrass made many of his mathematical discoveries while working as a village schoolteacher. By the time he did make a breakthrough into visibility in his mid-thirties he was already suffering from attacks of dizziness and sickness that would dog the rest of his life. Yet Weierstrass was to take the final step that overcame the calculus problem.

In the 1850s, he developed a formal definition of what was happening in calculus that was supposed to do away with the need even to bother with potential infinity (although the symbol ∞ somewhat confusingly continued to be used). Instead of worrying about the misuse of infinitely large and infinitely small quantities, Weierstrass proposed that the users of calculus thought about strict limits.

In Weierstrass's world there's no longer any need for variables, but rather a mechanism that will always enable us to crank the handle and come up with a suitably small (or large) value. Weierstrass's approach works by saying that a particular result is heading for a limit if, for any particular gap between the result and the limit, there was also a known gap for the factor driving the result. So, for instance, we no longer need to take that series

$$1 + \tfrac{1}{2} + \tfrac{1}{4} + \tfrac{1}{8} + \tfrac{1}{16} + \tfrac{1}{32} + \cdots$$

all the way to infinity to get the value of the sum to be 2. We can say that the limit of the series is 2 as long as the difference between the sum of the series and 2 is smaller than a particular value (Weierstrass called this difference ϵ, the Greek letter epsilon) given any particular difference between the point the series has reached and 0 (this difference he called δ, delta). As Ian Stewart says in his survey of mathematics *From Here to Infinity*:

> The introduction of potential versus actual infinity is a red herring: the whole problem can be formulated in purely finite terms.[76]

Weierstrass had not so much exorcised the ghost of the fluxion as photographed it and shown it to be someone dressed up in a sheet. Infinitesimals would remain these fake ghosts until the 1960s, when the possibilities of imagining the real thing were finally spotted. We will return to them in Chapter 16. But the considerations that were to make calculus respectable did not make the whole concept of infinity go away. The first steps that were taken by Galileo in his *Discourse on Two New Sciences* were to be given new life by another Italian.

10

PARADOXES OF THE INFINITE

I can't help it; – in spite of myself, infinity torments me.

Alfred de Musset, *L'Espoir en Dieu*

IT WAS ANOTHER MAN OF ITALIAN STOCK, born in 1781, largely unknown outside the mathematical world, who took over the torch from Galileo in highlighting the implications of the infinite and the paradoxes it generates. Just as Galileo's character Sagredo had encouraged a deviation into discussing infinity for the sake of it, rather than for the original intention of working out how matter was held together, so this man would contemplate infinity detached from any worries about the rights and wrongs of calculus. His name was Bolzano.

This surname belonged to one Bernard Placidus Johann Nepomuk Bolzano, born not in Naples or Florence or Rome but in Prague, where his father's business as an art dealer had taken him. Bolzano was no boy genius. Physically he was a weak child, by his own account moody and suffering from persistent headaches – not the sort of boy who becomes popular with his peers. In school he did nothing to stand out academically, though he did go on to the University of Prague, where he admits that philosophy proved 'almost as difficult as mathematics'. Not an ideal start for someone who was to explore the boundary between mathematics and philosophy that is formed by infinity.

Yet it seems that the challenge of difficulty stimulated the young Bolzano, pushing him into original patterns of thought, rather than leaving him to follow, sheep-like, the teaching at the university. He was marked out as something special. His ability to challenge accepted wisdom and come up with startling new ideas, a tendency that is often punished rather than rewarded, was in Bolzano's case a boon. In 1805, still only 24, he was awarded the chair of philosophy of religion. In the same year he was ordained a priest, and it was with this status, as a Christian philosopher rather than from any position of mathematical authority, that he would produce most of his important texts.

Most, but not all. For the consideration of infinity, Bolzano's significant work was *Paradoxien des Unendlichen*, written in retirement and only published after his death in 1848. This translates as *Paradoxes of the Infinite*. Here Bolzano considered directly the points that had concerned Galileo – the conflicting results that seem to emerge when infinity is studied. In fact, as Bolzano says right at the start of his book:

> Certainly most of the paradoxical statements encountered in the mathematical domain... are propositions which either immediately contain the idea of the infinite, or at least in some way or other depend upon that idea for their attempted proof.[77]

Bolzano looks at two possible approaches to infinity. One is simply the case of setting up a sequence of numbers, such as the whole numbers, and saying that as it can't conceivably be said to have a last term, it is inherently infinite – not finite. It is easy enough to show that the whole numbers do not have a point at which they stop. Let's give a name to that last number whatever it might be and call it 'ultimate'. Then what's wrong with ultimate + 1? Why is that not also a whole number?

The second approach to infinity, which he ascribes in *Paradoxes of the Infinite* to 'some philosophers... and notably in our day... Hegel and his followers', considers the 'true' infinity to be found only in God, the absolute. Those taking this approach,

Bolzano says, describe his first conception of infinity as the 'bad infinity'.

Although Hegel's form of infinity is reminiscent of the vague Augustinian infinity of God described in Chapter 3, Bolzano points out that it is, rather, the basis for a substandard infinity that merely reaches towards the absolute, but never reaches it. In *Paradoxes of the Infinite*, he calls this form of potential infinity

> a variable quantity knowing no limit to its growth (a definition adopted, as we shall soon see, even by many mathematicians)... always growing into the infinite and never reaching it.[78]

This is the sort of infinity we saw used to justify the oddities of calculus before Weierstrass freed it from the need to be this fluid variable. As far as Hegel and his colleagues were concerned, using this approach, there was no need for a real infinity beyond some unreachable absolute. Instead we deal with a variable quality that is as big as we need it to be (or often in calculus as small as we need it to be) without ever reaching the absolute, ultimate, truly infinite.

Bolzano argues, though, that there is something else, an infinity that doesn't have this 'whatever you need it to be' elasticity:

> In fact a truly infinite quantity (for example, the length of a straight line unbounded in either direction, meaning: the magnitude of the spatial entity containing all the points determined solely by their abstractly conceivable relation to two fixed points) does not by any means need to be variable, and in the adduced example it is in fact not variable. Conversely, it is quite possible for a quantity merely capable of being taken greater than we have already taken it, and of becoming larger than any pre-assigned (finite) quantity, nevertheless to remain at all times merely finite: which holds in particular of every numerical quantity 1, 2, 3, 4, . . .[79]

In other words, for Bolzano there could be a true infinity that wasn't a variable 'something' that was only bigger than anything you might specify. Such a true infinity was the result of joining two points together and extending that line in both directions without stopping. And what's more, he could separate off the demands of calculus, using a finite quantity without ever bothering with the slippery potential infinity. Here was both a deeper understanding of the nature of infinity and the basis on which Weierstrass could build his 'safe' infinity-free calculus.

Bolzano goes on in *Paradoxes of the Infinite* to look at the nature of infinite sets and series, and the paradoxes that arise from them – but the importance of the man in the development of our ideas of infinity is not this study of paradoxes but the fact that, after years of the solidification of the idea that infinity was just this virtual destination of a variable factor, Bolzano was prepared to speak up for a true, almost tangible, infinity. And it was on this point that his great successor Georg Cantor, who was to work not only with infinite sets but with bigger sets still, was to build the picture of the 'true' infinity that we still use today. Bolzano was Cantor's intellectual father.

Not only did Bolzano consider the simple sequence 1, 2, 3, 4, ..., he also looked at the harder to pin down concept of every single number (all the fractions) between 0 and 1. At first sight there's nothing daunting about this idea. We can easily imagine producing all the rational fractions, the ones in the form of a ratio like $\frac{1}{2}$ or $\frac{2}{3}$ by simply repeatedly dividing the line. But what about all the others – the irrational fractions? There are not only the relatively comprehensible algebraic fractions that form square roots and the like, but also the wildly unpredictable transcendentals like π. The range of fractions between 0 and 1 gapes open like a huge chasm – the closer you get to it, the more detail you see, the more you are aware just how little of the detail you can actually comprehend.

Bolzano found an equivalent to Galileo's observation that each square (1, 4, 9, etc.) could be allocated to one of the counting numbers, meaning that somehow the apparently different infinities of all the positive integers and of all the squares were in fact the same. In the case of all the fractions between 0 and 1, he showed that there was a one-to-one correspondence with the infinity of fractions between 0 and 2. He did this using a *function*, a mathematical black box that turns one thing into another.

Functions are widely used in mathematics and the sciences – and it's worth spending a moment considering them, because they sound much more complex (and hence off-putting) than they really are. A function is merely a set of rules for turning one number into another. The power of a function is that once set up, it can be used on any number you like – which is why you will find functions heavily used in spreadsheets on computers. In effect, a function *is* a mathematical computer, an imaginary box that turns an input number into an output.

A simple computer spreadsheet might have the basic price of products in one column, and the price including tax in a second column. To get from one column to the other, the computer follows the rule 'take what is in the first column and multiply it by a number'. For example, items for sale often carry a sales tax. Let's say this is at a rate of 17.5 per cent. So the spreadsheet would say 'take what is in the first column and multiply it by 1.175' which provides the original cost with an extra 17.5 per cent.

In mathematics, the second column, the derived column that includes the tax, is usually called $f(x)$, where x is the value of the first column. In other words, we are using a function, f, that starts with something, x, and converts it into something else, $f(x)$, defined by the inner workings of the function. So our sales tax function would give $f(x) = 1.175 \times x$. To get the result of the function we multiply whatever we are starting with by 1.175. This is *all* a function is – it's very valuable as a mathematical tool, but does cause a lot of confusion, partly because it sounds so grand, and partly because the way of describing the function's result as $f(x)$ looks mysterious – but that is the absolute convention, and confusing it will remain.

Arguably even more confusing is the similarity between $\times$ and x in the term $1.175 \times x$ in the previous paragraph. This rotated cross symbol $\times$ was first introduced as the sign for multiplication by the mathematician William Oughtred in his 1631 book *Clavis mathematicae.* Oughtred, a full-time clergyman, took private pupils in mathematics, including both John Wallis and Christopher Wren. His invention of an early circular slide rule at around the same time caused a furious argument with another of his pupils, Richard Delamain, who also claimed to have invented the device. But this was a more local dispute than the arguments that arose over the × symbol.

There was an outcry, particularly from Leibniz,[80] that this new multiplication sign was too similar to the letter *x*, and mathematicians and scientists often use a centred dot, e.g. $1.175 \cdot x$ to signify multiplying instead. Of course, the multiplication sign was therefore cunningly changed to one that gets confused with a decimal point. No one ever said mathematical notation was totally logical. Computer programmers tread the middle ground by using a star, hence $1.175 * x$, probably because there is no × symbol on the keyboard.

Yet another way of signifying multiplication is simply to leave the sign out altogether, making the function we described earlier $1.175x$, and perhaps this is the most effective approach. In this book, though, we will continue to use the × sign, because it makes the fact that we are undertaking the process of multiplication more apparent and it remains the symbol that most of us learn in school.

Many functions are much more complicated, messy creatures than my $1.175 \times x$, but Bolzano used the equally simple function $f(x) = 2 \times x$ − in other words, the function worked on any number by multiplying it by 2. Take the number 3 − put it through the black box of this function and it comes out as 6. Just as in the spreadsheet column, functions don't have to be used on a single number, they can work on a whole set of numbers − so Bolzano applied the function to all the numbers between 0 and 1, and found that each one then provided a unique number between 0 and 2. This process could also be reversed with the function $f(x) = \frac{1}{2} \times x$. For every single number between 0 and 2, this new function would generate an equivalent, half its value, between 0 and 1. This insight, which meant that the range from 0 to 1 could stand in for any other range of numbers (instead of $f(x) = 2 \times x$ or $\frac{1}{2} \times x$ use the appropriate function), was also to be highly significant to Cantor's work.

In fact, to be precise, Bolzano actually showed in his book that all the fractions between 0 and 5, and all those between 0 and 12, could be linked in this way. He explicitly comments that every value in one continuum belongs only to a single value in the other *and conversely*. This 'continuum' Bolzano refers to is merely a compact way of saying 'every value between'. So the continuum from 0 to 1 is every fraction between 0 and 1. The term captures the image of the continuous link between two points of a straight line, taking in every single point in between. Beware of the exposition I have seen in some books that merely says every value in 0 to 1 has an equivalent, specific value in the range 0 to 2. This could be true while still leaving lots of values in the 0 to 2 continuum that didn't have an equivalent in 0 to 1. The relationship has to work both ways to show that there is a one-to-one correspondence.

The retirement in which Bolzano wrote *Paradoxes of the Infinite* was not voluntary. Despite his increasing concentration on mathematics, Bolzano had held on to the chair of the professor of religion until 1820 without any feeling that his work was incompatible with his position. Then, when he was still only 39 years old, he was forced to retire. The exact circumstances of Bolzano being expelled from the university were the subject of mystery over the years, though it was clear that he did not go voluntarily. Some blamed the Jesuits, suggesting that this intellectual enforcer wing of the Catholic Church had charged him with heresy. Others suggested that he had made powerful enemies in the Church hierarchy who wanted him out of the way. For many years, Bolzano's expulsion was the subject of conspiracy theories comparable in complexity with those surrounding the death of J. F. Kennedy.

In fact, though, it seems according to D. A. Steele, the translator of Bolzano's work into English,[81] that it was not the Church at all that deposed Bolzano, but the Austro-Hungarian government in Vienna. This was a period in which continental universities were going through the uncomfortable change from teaching only from standard texts to allowing their professors more freedom in the way they put their subject across.

It's hard now to appreciate just how big a change this was. Historically, the role of a university teacher was to drum into his students the exact details of an accepted text – often a text dating back to ancient Greece – and to put across the set interpretations of that text. It's a scholarly approach that allows for little change in viewpoint over the years. The new approach, the one we would now take for granted, allowed each university teacher to choose his own texts with his own interpretation. Despite the fact that Prague was yet to adopt the new system, Bolzano *had* assembled his own course, and in doing so had ignored (and even criticized) the work of one Burgpfarrer Frint, an influential Viennese academic.

This error in office politics seems to have been part of the cause of Bolzano's downfall, bringing him in direct opposition to a man who had his hands on the levers of power in Austro-Hungarian academic circles. But Bolzano also made a more straightforward mistake in the eyes of the Austrian government. At a time when there was considered to be no possible conflict between a right-eous war and Christian beliefs, he preached a sermon to the university saying that there would be a time when the use of war, that absurd attempt to prove what was right through brute force, would be held in the same abhorrence as duelling. He even went so far as to suggest that the need to obey the state would not be a matter of absolute unthinking necessity, but rather one that would be tested against conscience and against legal rights.[82]

Bolzano's words might seem perfectly acceptable now, in effect defending the concept of conscientious objection and the need for the state to operate within the law, but even in the twenty-first century it is still not uncommon for politicians to become un-comfortable when members of the Church dare to comment on matters that politicians feel belong only to them. As recently as

July 2002, when the appointment of the new Archbishop of Canterbury, Rowan Williams, was announced, there was opposition from conservative individuals who did not like Williams's publicly voiced concerns about the possibility of taking military action against Iraq.

In the early 1800s, Bolzano's words were little short of treason. In an imperial order, signed on Christmas Eve 1819, Bolzano was dismissed from his position and reprimanded for his views in which he had 'grossly infringed his duties as a priest and a tutor and a citizen'. Although the Church was to find there was nothing wrong with Bolzano's faith and the way he expressed it, calling him an 'orthodox Catholic' who it was impossible to call a heretic, the State was impassive. For daring to tread on the toes of politicians, and to suggest that the cannon fodder should have a choice in obeying the call, Bolzano had to go.

He spent 28 years 'retired', working on a wide range of subjects from mathematics to linguistics, though often finding it difficult to publish because of his dismissal. Financially secure with the aid of one Frau Hoffman who supported him for twenty years, Bolzano died in 1848. He was not crushed by his dismissal, but it is difficult to believe he did not suffer from the authorities' opinion that his views were not safe for communication to the universities and the people.

Where others working in the same period and even later, such as Weierstrass and the German mathematician Bernhard Riemann concentrated on the way series and analytical tools like calculus could make use of infinity in order to safely reach results, Bolzano's glory was that he was prepared to contemplate infinity itself, and lead the way for another mathematician, Georg Cantor, who would put infinity squarely on the mathematical map, despite fierce opposition and the onset of madness.

First, though, it is necessary to take a step back from the infinite, to a theory that underlies the very nature of number itself. To understand infinity better, it proved necessary to pin down the much more worldly concept of a set.

11

SET IN STONE

They cannot scare me with their empty spaces
Between stars – on stars where no human race is
I have it in me so much nearer home
To scare myself with my own desert places.

Robert Frost, *A Further Range* (Desert Places)

BEFORE WE SOAR INTO THE MATHEMATICS of infinity we have to come plainly down to earth. The nature of infinity is irretrievably intertwined with a theory that lies at the very heart of mathematics (even though it was only formally developed at the end of the nineteenth century) – the discipline called set theory. Curiously, the prime inventor of set theory, the father of our modern understanding of infinity, is not a name that is on everyone's lips.

According to my dictionary of dates, one of the famous people born on 3 March is William Godwin (1756), father of the creator of Frankenstein, Mary Shelley. Another who shares that birthday is George Mortimore Pullman (1831) of railway carriage fame. Then there's Alexander Graham Bell (1847), who we all thought invented the telephone until in 2002 the US House of Representatives passed a resolution crediting the Italian inventor Antonio Meucci with demonstrating his teletrofono 26 years before Bell patented the phone. And we should not forget the great English orchestral conductor Sir Henry Wood (1869), who began the world-famous series of concerts, the Proms.

One name that isn't mentioned in the dictionary is Georg Cantor (Georg Ferdinand Ludwig Philipp to give him his full name), brought into the world on 3 March 1845. Yet this was the man who was to devise a fundamental component of mathematics, and who finally dragged infinity off its virtual pedestal and into full view. In the process he was to go mad.

Cantor's early life was engagingly cosmopolitan. He was born in St Petersburg in Russia, his father a Dane, the family moving on to the German city of Frankfurt when Cantor was 11. The boy was away from home a lot, initially at schools in Frankfurt and Darmstadt and then studying in Switzerland at the Zurich Polytechnic. From here he transferred to the illustrious Berlin University, the ideal place to follow his chosen career in mathematics. Yet he was to spend most of his adult life in the relative remoteness of Halle.

Halle is not the first German town that springs to anyone's mind, either as a sightseeing destination or as a great centre of learning. True, this was the birthplace of George Frideric Handel, and Halle maintained a name for music, but it was not the obvious destination for a top-notch academic. Cantor seems to have regarded Halle as a stepping stone to greatness at other universities – but the invitations to move on never came. He was to stay stuck on the stepping stone, never reaching the far bank. Yet it was from obscurity in Halle that Cantor would develop one of the essential foundations of mathematics. In the early 1870s, this unrecognized mathematician published a series of papers that would describe exactly what sets were and how they operated.

Where the mathematics of infinity can seem mind-bendingly unnatural, set theory is based on very simple concepts that we all grasp intuitively from an early age. Though it might not have been formalized until recently, set theory is inherently invoked whenever we count things, or simply use a number. It is such a fundamental part of mathematics that there have been occasional attempts to teach it in primary schools, but while we all unconsciously make use of set theory every day, the way it is presented mathematically can be off-putting – and it is rarely long before it once more drops out of the syllabus.

The trouble is that fundamentals might support everything else, but they aren't necessarily easy to explain – just because they are the essential building bricks of a discipline doesn't make them a good starting point. The same can be seen in physics – quantum theory and quantum electrodynamics provide the fundamental theory on which much of physics and chemistry is based, but it would be madness to start the exploration of science in a primary school at the quantum level. That doesn't make quantum theory any less important though, and similarly the inappropriateness of set theory as an early learning method does not undermine the importance of the theory itself.

In his preface to a student text on set theory, Seymour Lipschutz, Professor of Mathematics at Temple University, comments:

> The theory of sets is at the foundation of mathematics. Concepts in set theory, such as functions and relations, appear explicitly or implicitly in every branch of mathematics. These concepts also appear in many related fields such as computer science, the physical sciences, and engineering.[83]

In essence a set is a group of things. In principle these could be any unconnected objects, but, to be useful, sets are usually constructed of things that have something in common. Georg Cantor, who was the first to clearly describe the operation of sets said:

> By a set we are to understand any collection into a whole M of definite and separate objects m of our intuition or our thought.[84]

The word that is translated as 'set' is *menge* (Cantor was writing in German), which could also be translated as an aggregate – a joining together of objects. Perhaps the most important part of the quote is that last part 'of our intuition or our thought'. While the items in a set can be physically connected together – for instance the set of all the objects in my pocket – they can equally be linked simply by concept. A set is any grouping of things we can make mentally. The set of all apples in the world has no physical linkage: it's just everything we think of as an apple. Even more clearly, this applies to the set of everything that is spherical.

Though Cantor's definition is a bit wordy, it puts across what a set is as well as any other. Sets can encompass vast numbers or the very smallest quantities. Each and every thing in the world belongs to a huge number of different sets. Consider yourself for a moment. You are part of the set of living creatures on the Earth – one of billions in that case. You are also part of the smaller set of human beings on the planet. This set is entirely included in that first 'living creatures' set – so it is called a subset, a term from set theory that has become part of normal language.

Homing in, you are also part of the much smaller set of human beings with your name. And you are the sole member of a set of human beings with your name who were born at the exact time and place you were. (The parts that make up a set are usually referred to as members or elements, like members of a club, or the elements that make up the heating parts on an electric fire or form the constituents of chemistry.)

In our zoom in from the set of all living creatures to the set of 'you', each smaller set was part of the bigger one. But things don't have to be that way. Say your eyes are brown – you are then a member of the set of creatures with brown eyes, which includes some human beings, but not all of them – and equally well includes my dog. Similarly, you are a member of the set of all things with the same weight as you, which will include not only other people and animals, but also rocks and machines, neither of which fall into the 'all creatures' set.

The concept of sets comes more naturally in some languages than others. Where in English, sets are used very loosely, there is a much more formal and logical set approach in the structure of languages like Chinese. Descriptions normally begin with the largest set, then work down through the subsets. So, for example, an address will begin with the country, then the province, the city, the district, the street and the building. This logic extends to names too: hence the practice of putting the surname (defining the family clan set) before the personal name. Similarly to define the distinction between Chinese and British people, the Chinese approach would be to say Britain country people (*ying guo ren*) or China country people (*zhong guo ren*). Some of the older Chinese sets could be quite poetic – for example, there was a set of 'things that look like a fly when seen from a distance'.

In mathematics, the way sets overlap is often depicted visually using Venn diagrams. When first introduced to these pictures, I felt there was something missing. I understood what they showed, but was inclined to think 'and so what?' This reflects just how basic and common sense much of set theory is – it often verges on the irritatingly obvious. Yet once the basics are in place you can do much more with it.

The Venn diagram gives a visual indication of the way sets interact. So, for example, in the diagram below, the whole outer rectangle is our all-enclosing set (sometimes called a superset, one that encloses others) of all creatures. The left-hand circle is all human beings, and the right-hand circle creatures with blue eyes. They don't have to be circles, by the way; circles just happen to have been chosen. Nor does the relative size of the circles necessarily indicate anything. It was this arbitrariness of Venn diagrams that also irritated me as a child.

The rather more positive aspect of these diagrams is the way that they can illustrate several sets in action. Where the two circles overlap, we have a smaller area which is human beings with blue eyes. The two circles added together forms the set of creatures that are human beings or have blue eyes (or both). The left-hand circle with a bite out of it is human beings with other eye colours, while the right-hand circle with a bite out of it is creatures with blue eyes that aren't human beings. Finally there's the cut-out shape of the rest of the rectangle when the circles have

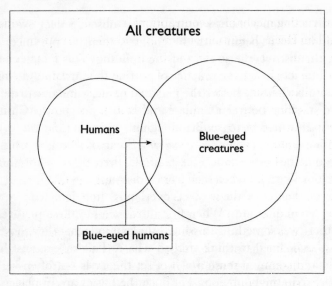

Figure 11.1 The set of all creatures.

been removed, which is every other creature that isn't a human being and doesn't have blue eyes. Quite an impressive amount of information for a little drawing.

Whether or not you like Venn diagrams, they are pretty well universally called that, even though there is considerable doubt as to whether or not John Venn, for whom they were named, deserves to be remembered by generation after generation of schoolchildren as the inventor of these graphical representations of sets.

Born in Hull in 1834, Venn spent most of his adult life in Cambridge University. Some modern writers have tended to play down Venn's abilities, yet at the time two of his books, *Logic of Chance* and *Symbolic Logic* were considered outstanding works. He was also anything but the typical, obsessive mathematician (which is perhaps why some more recent mathematicians don't like him). Not only did he write histories of both his college (Gonville and Caius) and the university itself, he delighted in constructing mechanical contraptions, rather like the drawings of William Heath Robinson in their inventive construction. Perhaps his greatest success on the mechanical front was a machine for bowling cricket balls. This was tried out against the visiting Australian cricket team in 1909, when to everyone's surprise (except possibly Venn's), the machine managed to bowl out one of the top Australian batsmen four times.[85]

It's certainly true that Venn produced for the first time the exact style of diagram we now use. However, the anti-Venn camp is swift to point out that the Swiss mathematician Leonhard Euler had already used a very similar diagram (fully overlapping circles that were concentric, where such contained circles tend to be off-centre in Venn diagrams, but which otherwise had much the same meaning) in his work *Opera Omnia*. Euler's diagrams aren't quite as flexible as Venn's – they are effectively a subset of the Venn diagram – but it's certainly true that Euler's name deserves to be there as much as Venn's. Even so, there does seem to be an element of sour grapes in the response of some to Venn's name being attached (not, as far as we can tell, at Venn's own request) to these diagrams. In his survey of mathematics, *The Mathematical Universe*, William Dunham writes:

> No one, not even John Venn's best friend, would argue that his underlying idea is very deep. Venn's innovation took immeasurably less brainpower than, for instance, Archimedes' determination of spherical surface . . . The latter required extraordinary insight; the former might just as well have been discovered by a child with a crayon . . . So, the Venn diagram is neither profound nor original. It is merely famous. Somehow within the realm of mathematics, John Venn's has become a household name. No one in the long history of mathematics ever became better known for less. There is really nothing more to be said.[86]

Venn diagrams (as I will continue to call them) also bring in that other bane of existence for those who were taught 'modern' mathematics – Boolean algebra. Its inventor George Boole sounds as if he ought to be French, but he was in fact born in the English city of Lincoln in 1815. Teaching himself in his spare time while working as an assistant teacher, Boole soon found that he was making original contributions, and after publishing some well-received papers, despite his lack of university education, he was awarded the chair of mathematics at Queen's College, Cork, where he was to remain for the rest of his life.

It was in 1854 that Boole published *An Investigation into the Laws of Thought, on which are founded the mathematical theories of logic and probabilities*, in which he described the first steps to a way of describing logical connections in a simple structured fashion – the method that would become known as Boolean algebra.

Boolean algebra is something that many of us now find more directly useful than Venn diagrams because of the advent of the Web, and specifically search engines. It uses key words such as 'AND' and 'OR' to combine different logical possibilities. Sets in the Venn diagram which overlap are combined using AND – in the example above, to be in the intersection of the two circles, a creature would have to be both human AND blue eyed. (The words don't have to be in capital letters, but this is often used as a convention to make it clear which are the special words or 'operators'.)

In set theory this AND connection is referred to as the intersection of the sets, reflecting the way that the circles in a Venn diagram intersect. Similarly, when we combine two shapes that don't totally overlap we are using OR for the combination. Creatures in the shape made by both circles combined in the diagram above are human OR blue-eyed. In set theory this is called the union of the sets human and blue-eyed, again describing the way the two circles are joined together, made one – a union. When we look at everything except the 'human being' circle, we are seeing the creatures which are NOT human beings – so the right-hand circle with a bite taken out of it, is creatures with blue eyes AND (that are) NOT human beings.

The same commands could be put into a Web search engine, if

you wanted information about, in the first case, blue-eyed humans, in the second, anything that was human or blue-eyed and, in the third, blue-eyed creatures that were not humans.

These magic words AND, OR and NOT provide the basic mechanisms for combining and manipulating sets. The whole business of sets had operated informally for thousands of years. We naturally classify things, generating sets based on pattern recognition. As we have already seen when exploring series, this practice of forming patterns is not just a handy habit, nor a matter of striving for artistic simplicity, but reflects the way the brain operates. Huge though the capacity of the brain might be, we can't handle every bit of information we need to cope with the world. Imagine if you had to deal with the whole world, atom by atom. Instead of saying 'there's George', you would have to analyse each atom and from the combined picture (which is changing all the time) recognize a person.

Similarly, as simple a task as switching on a light switch would take on huge proportions. Instead of asking someone to switch on 'that light switch' you would have to describe exactly where in space you want them to stand. Exactly which position each part of their hand would need to be positioned to. Exactly how much pressure to exert on the object with certain dimensions you find at a particular location. It could take all day. Instead, we make use of the set of things called 'light switch' – which can be quite different in appearance – to put across a huge amount of information in a short request.

Our brain does not hold a picture of every single light switch it has ever dealt with, which it flicks through, like a crime victim looking through a book of mugshots, every time it needs to deal with one. Instead it holds a very generalized, largely pictorial description – a mental model – of what a light switch is like. This is our mental set labelled 'light switches'. This is the pattern by which we recognize a switch.

So such pattern forming is very natural. We don't look at every dog we see and think 'a four-legged animal that's this size and colour and with these characteristics', we think 'a dog', meaning a member of the loose set 'dogs' (or more precisely 'things I recognize as a dog'). It is this pattern forming ability that makes us much better at recognizing things than a computer. The computer scientist faces huge challenges when attempting, for example, to get a computer to consistently recognize something as inconsistent as a face – seen from many directions and angles, ever changing in expression. Yet we, with our superb mental modelling skills can recognize faces so well that you can often take away one of the prime characteristics – for instance shave off a beard – and we will be hard pressed to recognize what it is that has changed.

Our ability to squeeze things into patterns and types is, therefore, extremely valuable – though it does have its downside. It means we have a natural urge to stereotype, pushing individuals into particular sets – the set of your nation, your occupation, your gender, your age band, your religion (or lack of it). It also means, as we saw earlier with stars in constellations, we can easily fall into the assumption that a pattern exists where none exists at all.

Almost all gamblers believe that after a run of (say) black numbers on a roulette wheel, it's time for a red one, because the pattern says there should be about the same number of each. The fact is, even if there had been ten black numbers in a row, the next number still has a 50 : 50 chance of being black or red – probability (and a fair roulette wheel) has no memory – but our insistence on seeing patterns makes us think differently.

In this particular instance, the pattern makes a certain outcome seem more likely. Let's take the simpler example of tossing a coin. If we toss a coin once, the chances are 50 : 50 of being heads or tails. So on average in half the cases (1 in 2) there will be head. Let's say in a particular toss it was a head. Then tossing a second time, again there's a 50 : 50 chance of getting a head. Probabilities combine by multiplying, so the chances of getting one head, then another are $\frac{1}{2} \times \frac{1}{2}$, i.e. $\frac{1}{4}$. If we continued to toss the coin for a total of ten throws and heads came up every time, the chances of this happening are

$$\tfrac{1}{2} \times \tfrac{1}{2} \times \tfrac{1}{2} \times \tfrac{1}{2} \times \tfrac{1}{2} \times \tfrac{1}{2} \times \tfrac{1}{2} \times \tfrac{1}{2} \times \tfrac{1}{2} \times \tfrac{1}{2} = 1/1024.$$

So there is a 1 in 1024 chance of throwing ten heads in a row. It's unlikely, but it could happen. Once it has happened, it is done. There is still a 50 : 50 chance that the next throw will be a head. We only see the sequence of ten heads as being significant because there is a recognizable pattern. If the sequence had instead been H T H H T H T T T H, we would not have seen it as particularly remarkable. But bear in mind that the chances of getting this *specific* sequence are also 1 in 1024. There is no difference in the probability, only in the perceived pattern.

Whether pattern recognition is operating for our benefit, as when we use it to recognize a person, an object or a word (again, computers are hard pressed to realize that THIS and this have the same meaning), or to our disadvantage when we imagine a significance that just isn't there, this pattern recognition could just as easily be called set recognition. We are identifying the set of characteristics that identify what a 'Brian Clegg's face' is, or the set of objects we can classify as 'a strawberry' when our brains construct these patterns. At the mental level, sets are an essential part of the way we interact with the world.

One reason that set theory is also described as being the fundamental basis of mathematics is that pretty well every type of mathematical operation, from addition and subtraction to the heavy-duty functions, can be defined from set theory. But another reason is the way sets are used to define the actual nature of number, without which there's little point to mathematics. I can have two completely different sets – for example, the set of 'the legs on my dog', and the set of 'the horsemen of the apocalypse'. There is no relationship at all between a collection of short, hairy limbs and those mythical anthropomorphized creatures, apart from the fact that both sets have the same number of elements – in this case four. This essential property of a set, the number of its elements, is called its *cardinality*.

It might seem that cardinality is just a fancy word for the size of a set, but it's not. We can establish whether two sets have the same cardinality even if we don't know how many things there are in the set – or even what a number is. In the examples above, if I put my dog alongside the horsemen of the apocalypse, I could pair the elements off – perhaps the left front leg with death, the right front leg with famine, the left back leg with pestilence, and the right back leg with war. At this point all the legs have been used up, and so have all the horsemen. Even if I don't know how many there are, I know that they have a one-to-one correspondence – they have the same cardinality.

If this process seems familiar, it shouldn't be too surprising. It is exactly the same approach taken when looking at Galileo's pairing off of the whole numbers and squares. Each number could be assigned a square. Each square can be assigned to a whole number. It doesn't matter that we don't know how many whole numbers there are, nor do we know how many squares – we do know that they have the same cardinality.

Generally speaking, we use numbers in two ways – to indicate the order in which objects were placed, and to indicate how many objects are present – the cardinal numbers which are these counting numbers, are the 'actual' number of objects in a set. Another mathematician, one Giuseppe Peano, devised rules that enabled the cardinality property of a set to give a definition for these cardinal numbers – the definition that is still the formal description of what a particular cardinal number is.

Peano was born on a farm near the town of Cuneo in the Italian region of Piemonte in 1858. His talents were recognized from an early age, and one of his uncles, a lawyer priest, took him to Turin to get a better education. He was to continue there to attend university, and it was also at the University of Turin that Peano would later teach. He had already produced several successful publications when in 1889 he wrote a pamphlet called *Arithmetices principia, nova methodo exposita*, a development of the counting numbers from set theory. In writing this extended paper, Peano quaintly adopted the ancient tradition of using Latin rather than a modern language.

There was a time when all academic communication was in Latin. When the universities were being established back in the twelfth and thirteenth centuries, the natural language for the scholars, just as it was for the Church, was Latin. It meant that a major university like Paris could take in students from all over Europe without any concern for their native tongues – all teaching could be in this common language.

It wasn't until the seventeenth century that much academic output reverted to the native tongues of the scholars, and even then, for example, Newton published some of his work, such as the great *Principia*, first in Latin – and even when he wrote a book like *Opticks* in English, it was later published in Latin translation (primarily for the European market). Until recently, there has been no challenger to Latin's place as a universal language. Certain languages have dominated in some areas – when my father was training as an industrial chemist, he was recommended to learn German, as so many chemical papers were then written in that language – but it is only recently that the combination of the dominance of the USA as a single superpower and the shared environment of the Internet has made English the inevitable choice for most academics, whatever their native language.

This difficulty of universal comprehension probably explains why Peano wrote his key work *Arithmetices principia* in Latin. He wanted to publish in a language that would be acceptable anywhere. His dedication to publishing in a universally readable format can be gathered from a venture he made in 1903. Peano had continued his success as a mathematician, but seemed to feel he wanted to leave more to posterity than his mathematical discoveries. He devised a new, artificial language, *Latino sine flexione*, which was essentially Latin stripped of its grammatical complexity with added vocabulary from Italian, English, German and French to bring it up to date.

Peano hoped that *Latino sine flexione* would become the new standard for publishing academic papers, restoring that universal accessibility that Latin had given to the medieval universities. He published the last edition of his biggest mathematical work *Formulario mathematico* in *Latino sine flexione*, but like so many other artificial languages, his invention never reached a critical mass of usage. Until Peano died in 1932 he still held out hopes that his language would become a standard – and it was perhaps his early ruminating on universal languages that brought him to write *Arithmetices principia, nova methodo exposita* in Latin.

It might seem strange that Peano should need to write this text that would enable numbers to be defined, but those familiar numbers we use all the time – 3 loaves of bread in my shopping, 12 months in the year, and so on, have to come from somewhere. They seem such a natural part of life that it's easy to think of them as real things, like the loaves I was counting. But in reality these numbers, the cardinal numbers, are just symbols we use to represent the cardinality of a set. I can't hold 15 in my hand, or paint a picture of 5 – I can paint five objects, I can hold a written symbol for 15, but the numbers themselves have no physical reality. Peano's definition lets us build those numbers in terms of a series of sets that are almost hauled up by their own bootstraps.

It is the starting point of this sequence that is the hardest to get your head around. Set theory relies on there being something called the empty set. A set of nothingness. A set that defines the absence of anything else. Although it may seem obvious, we start by defining zero as being this empty set. We can then produce the rest of the counting numbers by effectively adding a new set for every number, each enclosing the previous one like a set of containers. The cardinal number of a set is the number of sets that it contains. The number one is the set that contains the empty set. The number two is the set containing the empty set and the set that contains the empty set. And so on.

Sets are usually described mathematically using curly brackets, so for instance the set of the four horsemen of the apocalypse would be shown something like

{ War, Famine, Pestilence, Death }.

The empty set is given the special symbol ∅, so the numbers become:

0	∅	The empty set
1	{∅}	The set containing one set (the empty set)
2	{∅, {∅}}	The set containing two sets (the empty set, and the set containing the empty set)
3	{∅, {∅}, {∅, {∅}}}	The set containing three sets (the empty set, the set containing the empty set, and the set that contains the empty set and the set containing the empty set)

Cantor's set theory, as we have seen, lies at the heart of mathematics, yet not everyone was happy with it. As sets were further explored, it seemed impossible to use them without encountering paradoxes – and yet mathematicians had always prided themselves on having the only truly 'pure' discipline, one that should be inherently self-consistent and without conflict. In fact it was hard to see, some would argue, how anything that had built-in inconsistencies, as set theory would prove to have, could be considered mathematics at all.

Henri Poincaré, a French mathematician with an acerbic line in criticism who was a contemporary of Cantor's (he was born nine years later than Cantor in 1854), called set theory an illness that was afflicting mathematics, of which it would eventually be cured.[87] He thought it was a dead-end detour on the road to the ultimate development of mathematics. Yet more than a hundred years later, set theory has not been ousted from its position. There have been some attempts to rationalize the problems that arise from set theory's habit of throwing up paradoxes, but startlingly, not only has set theory resisted being corrected, it has been shown that it is impossible to ever totally untangle some of the problems that set theory generates.

The classic problem that faces those who want to base all of mathematics on set theory was discovered by Bertrand Russell at the beginning of the twentieth century. Born in the Welsh county of Monmouthshire in 1872, Russell was a ground-breaking philosopher who rather than deal with airy-fairy concepts of the nature of thought or whether matter exists independent of the observer, was determined to establish a logical philosophy underpinning the nature of mathematics. Like Bolzano before him, Russell was to lose his academic post because of his opposition to war. A life-long pacifist, Russell was dismissed from Trinity College, Cambridge, convicted twice for anti-war activities and sent to prison for six months in 1916.

Unlike Bolzano, though, Russell did not stay on working quietly in retirement. Although he would again be refused an academic position during the Second World War (this time when he was fired from City College, New York, for being morally unfit to teach there), he never failed to receive academic status, and was showered with accolades in later life. He received the Order of Merit, the British honour that is, in the words of the UK government royal website,

> given to such persons . . . as may have rendered exceptionally meritorious service in Our Crown Services or towards the advancement of the Arts, Learning, Literature, and Science or such other exceptional service as We are fit to recognise,[88]

and awarded the Nobel Prize for Literature in 1950 'in recognition of his varied and significant writings in which he champions humanitarian ideals and freedom of thought'.[89] But it was in his 1903 book *Principles of Mathematics* that Bertrand Russell was to stir up the supporters of set theory. Russell demonstrated that set theory is capable of generating uncomfortable truths. To do this, he made use of the initially convoluted concept of sets that have other sets as their members.

Think of the human race. Each one of us is a member both of the larger set of 'humanity' and the smaller set of 'our nation'. Humanity is a set that contains all of the sets of nations – as well as their subsets, such as people who live in a particular city. So the various sets of nations – American, French, Chinese, British, and so on – are all members of the set 'humanity'. So far, so good. Russell's next step is to introduce sets which have themselves as a member. We've looked at the set of 'humanity' – now think of the much larger set of 'everything that's not human'. This set includes itself – because 'the set of everything that's not human' isn't itself human. (The one assumption we are making here is presuming that a set is a thing, but Russell was more precise in his use of terminology.)

Now Russell has us ready to be confused. Some sets are members of themselves. We've just seen one. Similarly, some sets aren't members of themselves. The set of 'all pieces of music' is not a piece of music. A library – a set of books – is not itself a book. Russell took all the sets that are *not* members of themselves. Think of the whole set of these sets – the set of 'sets that aren't members of themselves'. Is, Russell asked, this set a member of itself?

You will probably need to reread that: is the set 'sets that aren't members of themselves' a member of itself? If it *is* a member, then it's a set that isn't a member of itself – so it *isn't* a member. If it *isn't* a member, then it's not a 'set that isn't a member of itself' – so it *is* a member after all.

This careful dissection of definitions shows a contradiction at the very heart of the nature of sets. It is because sets have this twist built into them that we are able to construct the familiar little word paradoxes, as when you try to work out whether or not it's true when someone says 'I'm lying'. Set theory seems to provide the perfect description of how mathematics works and how it interfaces with the real world of objects, yet it has this contradiction built into it.

Like all mathematical constructs, set theory depends on a set of 'axioms', fundamental assumptions that have to be made, and from which everything we know about a set can be deduced. Cantor's original ideas were slightly modified over time to become the accepted standard for set theory, called Zermelo–Fraenkel after the two mathematicians responsible for it. Yet even this system of eight basic axioms retains paradoxes, including the one Russell had pointed out. It seems impossible to understand mathematics without set theory, yet equally there appears to be no way to have set theory without the potential for confusion.

So it was that set theory would become an accepted part of mathematics, demonstrably right at the interface between mathematics and reality, yet still featuring those strange paradoxes. This is still the case today. Yet, powerful though it is, Cantor's development of set theory was only the beginning of his overthrow of the comfortable certainty of mathematics. Next he was to go for infinity itself. And more.

12

THINKING THE UNTHINKABLE

For suddenly it was clear to me that virtue in the creator is not the same as virtue in the creature. For the creator, if he should love his creature, would be loving only part of himself; but the creature, praising the creator, praises an infinity beyond himself.

Olaf Stapledon, *Star Maker*

GEORG CANTOR'S SET THEORY opened up a gateway. It was as if you opened a door and on the other side of it was the surface of the Sun. It was a remarkable achievement, but it carried danger along with it, as set theory made a close study of infinity possible. Cantor himself was to suffer for his dalliance with the infinite – and he was not to be the last great mathematician to have his mind damaged by the attempt to handle such an impossible concept.

It's easy to see this as overdramatization. Cantor worked on the infinite, certainly. Cantor suffered a series of nervous breakdowns, which is also a matter of record. But are the two really connected? It seems likely. The constant effort to encompass the unthinkable must have had an insidious impact on the man's mind. And we have the evidence of Cantor's own letters. After his first breakdown during 1884, from which it took him two months to recover, he wrote to a friend, Gösta Mittag-Leffler, saying that the last thing he had been working on before losing the ability to concentrate was in the field of infinity.[90]

His illness came back, more and more often. His hospitalizations were longer and longer. It was during one stay in the Nervenklinik (Halle's equivalent of a modern clinic where politicians and film stars might check in to recover from a breakdown) that he finally died on 6 January 1918. As the mathematics of infinity was Cantor's obsession, it is hardly surprising that he was working on it each time he suffered a breakdown, and yet it is equally difficult to avoid suggesting a connection between the subject and his condition.

When we use the term infinity in the crude way of English language, rather than with the scalpel-precise wording of mathematics, we tend to mean something like 'going on for ever' or 'the biggest thing that could possibly ever be, and a bit more'. Despite the triumphant battle cry of Buzz Lightyear in the Disney/Pixar animation *Toy Story*, 'To infinity and beyond!'[91] it is impossible with this normal linguistic version of infinity to imagine what going beyond infinity really means. It is almost a contradiction in terms – we are considering going further than something we define as the biggest thing that could be.

The same restriction holds true when we look at the spiritual contemplation of infinity. When religion links infinity with the Godhead it is saying that this is the ultimate limit. There is nothing more, nothing greater. Yet Georg Cantor was to contemplate going beyond infinity – and not just in a small way, but infinitely beyond.

This surprising outcome was a result of thinking about the different types of number and the arithmetic of infinity. For thousands of years infinity had been a convenient notion that was glimpsed from the corner of the eye. Look at it directly and it disappeared, but keep it marginalized and it made calculus and all of its dependent mathematics, physics and engineering possible. Cantor found a way to pin infinity down, to hold it in place.

To distinguish this 'real' infinity from the virtual fuzziness of potential infinity ∞, Cantor gave it a new symbol (it's a little more complex than that, but we'll come back to the naming in a little while). For the sort of infinity that contains all of the natural counting numbers he used the name $\aleph_0$, pronounced either 'aleph-zero' or the less prosaic sounding 'aleph-null'. Aleph is the first letter of the Hebrew alphabet, corresponding to A or the Greek alpha.

One of the reasons mathematicians had fought shy of envisioning the existence of such an infinity was the unnerving behaviour of aleph-null. As we get to know this remarkable non-number, it will prove to exhibit some strange properties. For instance, add 1 to $\aleph_0$ and you get $\aleph_0$. Similarly $\aleph_0 + \aleph_0 = \aleph_0$, and perhaps most strangely of all, $\aleph_0 \times \aleph_0 = \aleph_0$. Yet in reality these are behaviours that shouldn't seem quite so worrying. For instance, we don't have any real problem with accepting that $0 + 0 = 0$ or that $1 \times 1 = 1$. We already handle numbers that are special cases within the rules of arithmetic – why should the behaviour of aleph-null be any different?

This similarity of behaviour of infinity and 1 led Galileo to a rather confusing conclusion. As we have already seen, he correctly decided that the infinity of all counting numbers and the apparently smaller infinity of the squares was the same. In *Dialogues Concerning Two New Sciences*, his protagonist comments:

> In the preceding discussion we concluded that, in an infinite number, it is necessary that the squares and cubes should be as numerous as the totality of the natural numbers because both of these are as numerous as their roots, which constitute the totality of the natural numbers. Next we saw that the larger the numbers taken the more sparsely distributed were the squares, and still more sparsely the cubes; therefore it is clear that the larger the numbers to which we pass the farther we recede from the infinite number; hence it follows that, since this process carries us farther and farther from the end sought, if on turning back we shall find that any number can be said to be infinite, it must be unity. Here indeed are satisfied all those conditions which are requisite for an infinite number; I mean that unity contains in itself as many squares as there are cubes and natural numbers.[92]

Galileo's argument, though incorrect, is ingenious. He points out, rightly, that there are an infinite number of squares (or cubes) in the infinite range of the natural numbers, because as we have already seen each number has a square – there is a one-to-one correspondence from natural number to its square. Yet the strange paradox here is that this relationship isn't reversible. Although there *is* a natural number corresponding to each square, its root, there are also plenty of natural numbers that aren't squares. And, because of the nature of multiplication, the bigger the number, the fewer the squares there are. There are two squares in the first nine numbers (4 and 9), but only seven in the first 64 (4, 9, 16, 25, 36, 49 and 64).

So, Galileo reckons, as you head to bigger and bigger numbers you are somehow getting farther away from infinity rather than closer to it. However, when you look at the (rather limited) set of number that is unity, you find that there is, just as with infinity, a one-to-one correspondence between the natural numbers (1), squares (1×1) and cubes ($1 \times 1 \times 1$). They're all 1. The error Galileo seems to fall into is assuming that two things with similar properties are the same – rather like, for instance, having decided that everything with white curly fur and four legs is a sheep, deducing that both a white Scotty dog and some white sofas are also sheep. Even so, there's something rather attractive in Galileo's powerfully imagined concept – in some senses unity, oneness has got strong parallels with boundless infinity.

However, Cantor's world of aleph-null is capable of much more strangeness than the familiar number 1 can conjure up. We ought first of all to see how he coped with the aspect that caused Galileo such trouble, and that only happens to apply in the case of 1 because it is equal to its own square. How is it possible for the whole set of numbers in infinity (for example, all the counting numbers) to have a one-to-one correspondence with a subset of those numbers (for example, the squares)?

Cantor overcame this apparent impossibility by building it into his definition of the infinite. He *defined* an infinite set as one that had a one-to-one correspondence with a subset. And the problem disappears.

In his definition, Cantor was saying that an infinite set will have the same cardinality as a subset of itself. We tend to think of cardinality as the number of items in a set – so when we say that there are four eggs in a basket we mean that the set of 'the eggs in this basket' contains four elements. Strictly though, as we saw in the last chapter, cardinality is about having one-to-one correspondence with another set, usually with the set that defines the natural numbers from Peano's axioms. So though you may say, for instance, that there are less even numbers than there are integers, they have the same cardinality, as we can match off each even number against an integer (showing that the set of the integers is an infinite set, as it has the same cardinality as the even numbers, a subset).

Numerical sophistry? Not at all. You have to remember that mathematics is not based on the real world, but on a set of carefully constructed rules. As long as you follow those rules, you can work through the mathematics. You may be doing this for the sheer hell of it (as many theoretical mathematicians are), or it may be that those mathematical constructs can be extremely valuable in dealing with the real world.

This real value can even be present when the mathematics itself has no real-world basis. After all, vast tracts of physics and engineering depend on using a special kind of number called an imaginary number that has no corresponding real-world concept. Imaginary numbers date back in a crude form to the sixteenth century and reflect a simple problem that arises from the fact that multiplying two positive numbers together provides a positive number (for instance $3 \times 2 = 6$), and multiplying two negative numbers together *also* gives a positive result (so $-3 \times -2 = 6$).

That in itself isn't a problem until you start thinking about square roots. It's easy enough to see that $\sqrt{4}$, the square root of 4, is 2 (or possibly -2) because $2 \times 2 = 4$ and $-2 \times -2 = 4$. But what is $\sqrt{-4}$? What number multiplied by itself produces a negative number? It can't be a positive number – the result would be positive. And it can't be negative – the number would be positive again. If there is such a thing as the square root of a negative number, it must be something else, something special. The square root of -1 was given the symbol i, making $\sqrt{-4}$ become $2i$, twice the square root of -1.

The basic idea of working with the square root of a negative number emerged from a melting pot of Italian mathematical striving for ascendancy in the sixteenth century. Three men would each take a step towards the realization of the significance of these numbers. The first was Niccolò Fontana, born at Brescia in 1499. Fontana was horribly wounded as a 12-year-old when the French occupied his home town, receiving a terrible slash from a sabre across his face. He was expected to die, and when he did recover the damage to his mouth and palate made it difficult for him to speak. He was given the nickname Tartaglia, 'stammerer', which stuck with him for life.

From a humble background, Tartaglia had no university education, but taught himself enough mathematics to enable a natural talent to shine through and to keep him employed as a mathematics teacher. He emerged into the limelight in a public mathematical duel with another mathematician, named Fior, over the solution of cubic equations – equations like $x^3 = 1$, a topic that was all the rage at the time. Tartaglia clearly had a method for solving these equations as he trounced Fior, but would not say how he did it.

Tartaglia's contemporary Girolamo Cardano (sometimes called Jerome Cardan), a mathematician with much greater academic standing, was fascinated by Tartaglia's ability and asked if he could use his method in a book he was writing. Tartaglia refused to allow it, and here the trouble really began. Cardano tried several times to persuade Tartaglia to describe his method and finally resorted to bribery, hinting that he could get Tartaglia a position with a powerful military governor. Tartaglia gave in and described his method in confidence, in return for a promise from Cardano that it would never be revealed.

In the next few months, Cardano managed to extend Tartaglia's formula to make it more general, and soon after, much to Tartaglia's horror, published it in his book *Artis magnae sive de regulis algebraicis liber unus*. This wasn't plagiarism, Tartaglia was fully credited in the book, but Cardano had broken his word, and Tartaglia never forgave him and went on to attack Cardano in word and in print.

In Cardano's book, one consequence of the work on cubic equations was a short description of a solution involving the square root of a negative number, though Cardano describes this as being 'as subtle as it is useless'.[93] The idea was strengthened by another Italian, Rafael Bombelli, working at much the same time, who was the first to think of negative numbers as separate entities (rather than merely considering the subtraction of one positive number from another). Bombelli recognized that there could be something called -3, and with this idea in place also specifically described the working of the square root of a negative number.

This is fascinating stuff, but the realization that brought i into the toolkit of physicists and engineers was the realization that you could use imaginary numbers to extend the number line into a new dimension. We have already met the idea of a ruler-like line stretching off to the unimaginably large negative at the left, passing through 0 in the middle and heading off towards infinity at the right. Now put a second line at right angles to this, passing through 0, like the second axis of a graph. Call this the imaginary number line. As you move upwards from 0 you are going through bigger and bigger imaginary numbers: i, $2i$, etc. Below the conventional number line you have negative imaginary numbers. This was the contribution to the development of imaginary numbers made by the nineteenth-century German mathematician Carl Friedrich Gauss, and it was the final step needed to transform imaginary numbers into the powerful tools they are today.

Now instead of a number line you have a number plane (that's a flat surface, rather than an aircraft), a whole two-dimensional world. Just as a point on the number line defines an ordinary number, a point on the number plane defines a complex number – a combination of a real number and an imaginary number. So a point two units along the real number line and three units up the imaginary number line would be $2 + 3i$. It is these complex numbers, and the mathematics developed around them, that work so well in handling many of the problems of the real world. They make it possible to solve a whole range of equations that had been impossible to address up until then – the simplest being $x^2 = -1$ and which earlier mathematicians had simply labelled insoluble. The complex numbers work perfectly when describing anything involving waves, where equations of this type abound. They are found anywhere from simple electrical work right through to the wave aspects of quantum mechanics.

Imaginary numbers, then, which are of immense practical use, are based on just as arbitrary a theoretical concept as an infinite set. It would be wrong to say 'imaginary numbers don't exist' or that they are meaningless. They have a very real meaning and existence within the rules of mathematics. But, just like Cantor's aleph-null, you cannot illustrate them directly with a real-world equivalent in the same way that you can illustrate the concept of a 'half' by slicing a cake into two equal parts. So we should not be in a hurry to dismiss Cantor's construct of a clearly defined infinite set even if the basis for the definition seems impossible in the 'real' world.

Defining the infinite wasn't enough for Cantor, though. The term aleph-null contains two elements. The aleph is there as the first letter of the Jewish alphabet. Why aleph was chosen is subject to some dispute. Amir Aczel, in *The Mystery of the Aleph*,[94] suggests that Cantor's family, though Christian, had Jewish roots, and that he would have been aware that in the Jewish mystical tradition, the cabbala, one of the representations of God as the infinite was aleph.

Cantor was certainly strongly influenced by religious considerations and would also use the Greek letter omega as a symbol for a different aspect of infinity. God is not infrequently referred to as 'the alpha and the omega' (the first and the last, the beginning and the end, as these are the first and last letters of the Greek alphabet) ever since St John wrote in his Revelation:

'I am the Alpha and the Omega', says the Lord God, who is and who was and who is to come...,[95]

and it may well have simply been as an alternative to the rather overused alpha that Cantor went for aleph.

And why the 'null'? Almost unbelievably, Cantor proved that there were bigger numbers than infinity. That $\aleph_0$ was only the starting point for greater infinities – and all this despite the fact that $\aleph_0 + \aleph_0 = \aleph_0$ and that $\aleph_0 \times \aleph_0 = \aleph_0$.

We will rediscover these supercardinals in Chapter 14, but first we need to put things in order.

13

ORDER VERSUS THE CARDINALS

And, as their wealth increaseth, so enclose
Infinite riches in a little room

Christopher Marlowe, *The Jew of Malta*

ALTHOUGH THE STRANGE BEHAVIOUR of aleph-null takes a little getting used to, once we see that $\aleph_0 + \aleph_0 = \aleph_0$ and $\aleph_0 \times \aleph_0 = \aleph_0$, it seems to make a kind of sense. However, infinity is one of those topics that refuses to come quietly. Just when you think you have got it straight, another twist is added into the equation. When Cantor devised aleph-null he was thinking of cardinal numbers – the number of elements in a set – yet the number system has one more meaning that he also had to accommodate in his scheme, which would lead to yet another type of infinity.

We saw earlier that a simple, familiar symbol like 7 could represent both the integer 7 and the property of the appropriate set that we call its cardinality. However, the symbol 7 has a third meaning. Normally we integrate these different meanings of numbers in our heads so firmly that we can find it hard to make the distinction, but as well as being an integer and a cardinal, 7 is also an ordinal.

There is a religious sounding feeling to the distinction between cardinals and ordinals, but the reality is much more mundane. The clue to the meaning here is in the sound of the word. An ordinal number is one dependent on order. If you imagine a row of items, the ordinal number 7 refers to the position of the item in the row − it's the seventh item. Imagine a group of ten people arranged in order of height. The cardinality of that group is clearly 10. It turns out that the person with six people smaller than him is called Fred. We can only meaningfully refer to the cardinality of Fred as 1, but we can say that his ordinal number is 7. It defines his position in the row.

In set theory, when you put the elements of the set into order it is (not surprisingly) called an ordered set. A very familiar ordered set is the non-negative integers, counting up in the sequence:

$$0, \ 1, \ 2, \ 3, \ 4, \ 5, \ldots$$

For a set to have ordinal values it has to be an ordered set, like this one. Note that order doesn't necessarily mean having a simple linear order. For example, $1, 3, 5, 7, \ldots$ is an ordered set, and so is $1, 2, 5, 14$. Any set where we can define a mechanism, however convoluted, for relationships in the set that defines where to put them has order. For instance, $\{5, 4, 1, 3, 2\}$ is in fact an ordered set, but the order is dependent on the spelling of the names of the numbers.

To have ordinal values, though, a set also needs one further quality − it needs to be what mathematicians call a 'well-ordered' set. This adds the requirement that every subset of the set has a first element. At first sight this is a crazy distinction. Surely all sets, once they have been put into order, have a first element? But in fact there's a very familiar set that isn't well-ordered − the integers. Given any set of integers, we can put them into order:

$$-5, \ -4, \ -3, \ -2, \ -1, \ 0, \ 1, \ 2, \ 3,$$

for example. But take in the whole set of integers, positive and negative, and where does it start? What is the first element of the subset 'the negative integers'?

We don't have this problem with the positive integers, so these are well-ordered and we can define ordinal numbers to match any particular set of them. Generally speaking, you can think of an ordinal number as being defined by the set that comes before it, so, in the non-negative integers, ordinal 5 is defined as $\{0, 1, 2, 3, 4\}$.

Cantor, of course, was not happy to limit his ordinals to the finite. But although with finite numbers there is no need for a distinction in symbol between cardinals and ordinals, once we pass the boundary into transfinite numbers there is a need. Remember that cardinality is defined by one-to-one correspondence, making it reasonable for $\aleph_0 + 1$ to be the same as $\aleph_0$. Things are rather different with the ordinals. Here, having reached infinity, there still has to be a 'next one'. And so for the ordinals a new symbol for the 'smallest infinity' was devised.

It's rather a shame there wasn't any consistency in the way the symbols were used, but this time, the smallest ordinal infinity was represented by ω, the lower case Greek letter omega, the last of the letters, and hence often used to represent the ultimate (as we have already seen, 'the Alpha and the Omega', the beginning and the end, is used in the Bible when describing God to emphasize his limitlessness). So ω is the limit of the sequence $0, 1, 2, 3, \ldots$, the first transfinite ordinal, but it is then followed by $\omega + 1$, $\omega + 2$, and so on.

One of the reasons we tend to confuse ordinals and cardinals (apart from using the same symbol for both) is that with ordinary finite numbers they are in effect the same. The ordinal number of a finite set will be the same as its cardinal number. But once we get to infinity and beyond (the various flavours of infinity are called transfinite numbers) we need to make the distinction, because it is possible for two sets to have the same cardinality, but different ordinals.

To get your mind around this concept, consider a set of numbers (it doesn't matter what they are) that go $\{a_1, a_2, a_3, \ldots\}$; the size of these a numbers doesn't matter, but they are in order. This infinite set has the same cardinality as another infinite set that goes $\{a_2, a_3, \ldots, a_1\}$, because we can go through both sets putting members in a one-to-one correspondence. But the second set has a different ordinality. The ordinal number of the first set is ω. This is also the ordinal number of $\{a_2, a_3, \ldots\}$ – but then we've the additional a_1, so the second set has the ordinal number $\omega + 1$.

The apparent sleight of hand in the difference between $\{a_1, a_2, a_3, \ldots\}$ and $\{a_2, a_3, \ldots, a_1\}$ means that the arithmetic of infinite ordinals is not as we would expect. This is because it is almost impossible not to fall into the trap of mentally equating cardinals and ordinals. The outcome of the slippery nature of ω is that $\omega + 2$ is not the same as $2 + \omega$, and $2 \times \omega$ is not the same as $\omega \times 2$.

The detail of this is a little messy, but let's just look at the addition aspect. We have seen elsewhere how we can say two infinite sets are equivalent, that they have the same cardinality, if there is a one-to-one correspondence of elements between the two. When you have a well-ordered set, you can't establish a one-to-one correspondence between the set and any initial segment – a piece chopped off the front of the set. If you have two well-ordered sets and one has this one-to-one correspondence with the first part of the other, the first set is said to be shorter.

Now, according to the rules of how sets combine, adding together $\omega + 2$ comes up with the set $\{1, 2, 3, \ldots; ?_1, ?_2\}$, where $?_1$ and $?_2$ are the first two ordinal numbers after ω, while $2 + \omega$ produces the set $\{?_1, ?_2; 1, 2, 3 \ldots\}$. In the second case we can't generate an ordered one-to-one correspondence with any initial segment – we have to take the whole set – so $2 + \omega = \omega$, but in the first case there's a one-to-one correspondence with the first segment of the set, so the result is bigger than ω.

This means of combining results produces infinite ordinals which go up through an increasing scale $\omega + 1$, $\omega + 2$, and so on, until they reach $\omega \times 2$, which is the ordinal of the set $\{1, 2, 3, \ldots; \omega, \omega + 1, \omega + 2, \ldots\}$. This whole process moves on in a way that is reminiscent of the Archimedes' hierarchy of numbers in the *Sand-reckoner*. After working through $\omega \times 3$,

$\omega \times 4$, and so on, we get another infinite series reaching $\omega \times \omega$ or ω^2. Carrying on in this fashion we can reach ω^ω, and from there $(\omega^\omega)^\omega$, and so on. The process is imagined being repeated over and over until the sequence of ω powers stretches up like an infinite kite-string, reaching the ω position in that sequence. The new ordinal is given the label ϵ_0 as a new starting point. Here we start all over again with $\epsilon_0 + 1$, $\epsilon_0 + 2$, and so on, forever.

Cantor's infinite ordinals showed that it was possible to go on in a structured, an ordered, way beyond infinity when we are considering order. But what of the cardinals? Given the ease with which one-to-one correspondences could be established, surely aleph-null was as big as they came? In fact, though, not only was there a bigger cardinal infinity than $\aleph_0$, but it could be fitted into the apparently tiny gap between 0 and 1.

14

AN INFINITY OF INFINITIES

My bounty is as boundless as the sea,
My love as deep; the more I give to thee,
The more I have, for both are infinite.

William Shakespeare, *Romeo and Juliet*, II. ii. 133

ALEPH-NULL, $\aleph_0$, DESCRIBES THE SET of the counting numbers and also the squares, as we have already seen. Such sets are called denumerable or countably infinite. This sounds something of an oxymoron. If something is infinite, by definition it can't be counted. (How can you count something that has the same cardinality as a subset of itself?) What is meant, though, is that it is possible to number off each item in the set, to pair it off with a counting number. You will never reach an end, but the *process* is possible.

We know that this is possible with the set of even or odd numbers, the integers and squares of integers, but what about the rational numbers? Despite the clumsy terminology that seems to suggest these are numbers capable of reasoning, remember that these are simply the numbers generated from the ratio of two counting numbers. There are clearly a huge number of these. Just take the range of fractions between 0 and 1. We have an infinity of fractions like $\frac{1}{1}, \frac{1}{2}, \frac{1}{3}, \frac{1}{4}, \ldots$, plus another infinity of fractions with 2 on top, and so on. For an infinite number of times.

So could we have hit on this bigger infinity than $\aleph_0$ that Cantor seems to have been looking for?

No. Remarkably, even though there seem to be vastly more rational numbers than counting numbers, just as we did with the squares it is possible to set up a one-to-one correspondence between these ratios and the counting numbers. Cantor's demonstration of this takes a moment to absorb, but is purely visual – there are no equations to puzzle over. He arranged all the rational numbers in a number square, the sort of thing our children use to perform basic arithmetic. From left to right the top number of the ratio gets bigger, from top to bottom the bottom number of the ratio gets bigger. This infinite square would in theory include every single rational number. A small corner of the square is shown in Figure 14.1.

There's a lot of repetition in there. For example, the main diagonal, highlighted in the square, consists of the number 1 all the way through, an infinite number of times. But this doesn't matter. What Cantor then did was to provide a way of working through the entire table in a sequence. Each step through the sequence can be assigned to one of the counting numbers – so despite their apparent hugeness, the rationals have the same cardinality of $\aleph_0$ as the counting numbers. Cantor's route

1/1	2/1	3/1	4/1	5/1	6/1	7/1	8/1	9/1	10/1	...
1/2	2/2	3/2	4/2	5/2	6/2	7/2	8/2	9/2	10/2	...
1/3	2/3	3/3	4/3	5/3	6/3	7/3	8/3	9/3	10/3	...
1/4	2/4	3/4	4/4	5/4	6/4	7/4	8/4	9/4	10/4	...
1/5	2/5	3/5	4/5	5/5	6/5	7/5	8/5	9/5	10/5	...
1/6	2/6	3/6	4/6	5/6	6/6	7/6	8/6	9/6	10/6	...
1/7	2/7	3/7	4/7	5/7	6/7	7/7	8/7	9/7	10/7	...
1/8	2/8	3/8	4/8	5/8	6/8	7/8	8/8	9/8	10/8	...
1/9	2/9	3/9	4/9	5/9	6/9	7/9	8/9	9/9	10/9	...
1/10	2/10	3/10	4/10	5/10	6/10	7/10	8/10	9/10	10/10	...
...	...	...	...	...	...	...	...	...	...	...

Figure 14.1 Number square of rational numbers.

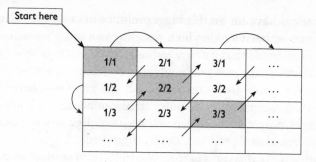

Figure 14.2 Corner of number square with Cantor's diagonal sequence.

through the square, just dealing with the first few numbers, is illustrated in Figure 14.2.

Cantor happened to use a diagonal sequence, but the actual sequence really wasn't important. He could equally have stepped through the square by the route shown in Figure 14.3.

The fact remains that each step can be counted, one at a time, to produce exactly the same one-to-one correspondence as we can between $\aleph_0$ and $\aleph_0 + 1$ or between $\aleph_0$ and $\aleph_0 \times \aleph_0$. The rational numbers also form a set with cardinality aleph-null, exactly the same cardinality as the counting numbers.

Most modern mathematical proofs are nightmares for the uninitiated. They can run to hundreds of dense, impenetrable

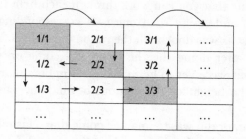

Figure 14.3 Alternative route through the square.

pages of calculation. Yet Cantor's proof is so simple and elegant that it can be put across without resorting to a single equation. In fact it is so simple that it's easy to think that it is one of those trite mechanisms that don't really advance anything. Because surely we can do exactly the same thing with any set of numbers? And that would mean that all infinities are the same. Yet with another blindingly simple proof that needs no painful equations, Cantor was to show that $\aleph_0$ was just the starting point.

The source of this superinfinity was the type of number that had given the Pythagoreans such a hard time in ancient Greece. Irrational numbers. Those numbers that could only be accurately represented as an infinitely long series of decimal fractions, never settling into a repeating pattern. In a way it makes sense that the irrational numbers should be responsible for pushing mathematics beyond infinity, because in one sense at least, each of these numbers is infinite in its own right – infinitely extended in value – so we are dealing with a set of numbers each of which already has an infinite property. But that's no proof. In 1873, however, Cantor found a way to work through a sequence of these numbers just as he had in the number square of the rationals – and the result, often called 'the diagonal argument', was very different.

Cantor imagined a sequence of every number (rationals and irrationals) between 0 and 1.

Figure 14.4 is a simpler version of the table of rational numbers. Here there is only one, infinitely tall column. At first sight you might assume that you can work through each item in the table one by one and assign each one to a corresponding counting number, just as was done with the rational numbers. Yet Cantor found that after doing this, he could still come up with extra numbers that were yet to be matched. Numbers that, in effect, increased the height of the column beyond the basic infinite cardinality.

What Cantor did was to imagine going down his table, taking the first digit from the first number, the second digit from the second number, the third digit from the third number, and so

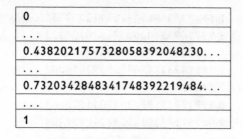

Figure 14.4 All numbers between 0 and 1.

on, all the way through the whole sequence. Next he changed every digit in this new number – for example, he could add 1 to every digit (shifting 9 up to 0). Now there's something very strange about the new number that Cantor had now produced. It is different from every single number in the table. Remember, its first digit had been the same as the first digit of the first number in the list. Now it is one more than that – so the new number couldn't be the same as the first number in the list. The second digit had been the same as the second digit of the second number in the list. Now it is one more than that – so the new number couldn't be the same as the second number in the list. And so on.

To make that clearer, let's imagine that the table had been scrambled before doing the digit by digit matching (it doesn't make any difference to the process, but it makes it much easier to draw out on paper, because if the table was in increasing order, it would start with an infinite number of values, each of which had 0 in as many digits as I could write out). My scrambled table happens to start with the numbers given in Figure 14.5.

I now pick out the first digit of the first number, the second of the second, and so forth – the digits in bold below. That gives me 0.6381392... We now add one to each digit after the decimal place to get 0.7492403... This can't be the same as the first number in the table because it has a 7 instead of a 6 in the first place after the decimal. It can't be the same as the second

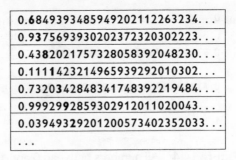

Figure 14.5 Compare scrambled numbers with 0.7492403 . . .

number, because it has a 4 instead of a 3 in the second decimal place. And so on. Forever.

It's an unavoidable fact – we have just proved that you can't make a one-to-one matching of every single number, rational and irrational, between 0 and 1 with the counting numbers. After we've gone through the whole column, there are still numbers left over. There are more numbers between 0 and 1 than there are integers. More than infinity. It's worth taking a moment to let that sink in. It's a remarkable concept. It feels uncomfortably strange, something that Cantor was well aware of. In fact, to avoid excessive fuss he gave his first paper on the subject a harmless sounding title 'On a property of the collection of all real algebraic numbers' to try to avoid drawing excessive attention and criticism to the subject before it was published.

Once Cantor had himself absorbed the implications that there were different kinds of infinity, he was curious to see how different mathematical constructs could be seen in the light of these different infinities. There were some similarities between the extra levels of infinity and adding new dimensions. We are used to the fact that when we draw on a piece of paper we are working in two dimensions – height and width, while in the real world we live in three dimensions (or four, if you want to add time into the equation). Mathematicians are comfortable with entities that exist in many more dimensional states, from the point (with no dimensions at all) and the line (a single dimension), through to the ten-plus dimensions that physicists now believe might be necessary to explain the behaviour of the cosmos, off to as many different dimensions as you like.

Infinity, it seemed to Cantor, also had different levels – not strictly dimensions, but a similar hierarchy, with the 'basic' infinity of the integers or rational numbers at the bottom as $\aleph_0$ – in effect, the minimal dimensioned infinity, aleph-null. The next step up the hierarchy, $\aleph_1$, he imagined taking in all the $\aleph_0$ values, so where $\aleph_0$ was the cardinality of the integers, the even numbers, the squares, and so on, $\aleph_1$ was the cardinality of the $\aleph_0$ types of transfinite number, and so on.

He had now also shown that the cardinality of the numbers between 0 and 1 fitted into a larger 'dimension' than $\aleph_0$, presumably one of the greater alephs (or even something in between levels), but he had not proved that it had the value $\aleph_1$. This did, however, seem likely, and the theory that the cardinality of these rational and irrational numbers was $\aleph_1$ proved useful enough to be given a name in its own right. It is called 'the continuum hypothesis' because it deals with the continuum of numbers between 0 and 1.

It's tempting to think that Cantor had some sort of picture of the different levels of infinity as dimensions, because he was soon to look at the impact of dimensions on infinity, to produce a result that he seems to have found even more staggering than the discovery that there was more than one kind of infinity in the first place.

When Cantor had shown that all the numbers between 0 and 1 provided a 'next generation' infinity, he had, in effect, explored the infinity of the points on a line. Just as Bolzano had, his results could be expanded to cover the range between 0 and 2, or 0 and any number. Cantor's numbers between 0 and 1 were, in effect, the number line from 0 to 1, and so provided a description of one-dimensional infinity. But what was the infinity of a piece of paper? Or a sugar cube? Having found one extra level of infinity, Cantor seemed enthusiastic to add more to his worldview.

He started by looking at a very simple two dimensional shape – a square, one unit per side. With the shape in place, he had to find some way to represent every point on the square. Luckily, this was a problem that had been sorted out many years before. The man largely responsible was the French polymath René Descartes.

Descartes was born in the small town of La Haye in the Touraine region of France in 1596. From the age of eight, this son of an aristocratic family attended the La Flèche Jesuit school in Anjou. From there he went on to study law at the University of Poitiers, but this he intended to be a mere stepping-stone to a military career. Soon after leaving university he entered the service of Prince Maurice, the ruler of the Netherlands.

The realities of army life proved less attractive than the promise. Descartes soon came home to France, where perhaps inspired by his Jesuitical training, he began to take a serious interest in natural philosophy. For eight years after his return he regularly travelled in Europe, spending time in Bohemia, Hungary and Germany. But while he had not felt any urge to stay with Prince Maurice, he had fallen for the Netherlands as an attractive place to live. Descartes returned there when he was 32 and stayed for most of the rest of his life. In 1637 he published at Leiden a work called *Discours de la méthod pour bien conduire sa raison et chercher la vérité dans les sciences*. This discussion of scientific method, though wordy, is of interest in its own right, but the most significant parts were three appendices: on optics, meteorology and mathematics.

In the third of these, the *Géométrie*, it is not the way Descartes deals with geometry in isolation that is interesting, but rather the way he combines algebra and geometry, setting up a relationship between the spatial mathematics of geometry and the formulas of algebra. The potential for linking the two was nothing new, but what Descartes did in the *Géométrie* was to provide a simple, consistent mechanism from translating from equations into lines in space. This was to become termed analytical geometry.

Descartes divided two-dimensional space into an arbitrary 'up and down' and 'left and right' split. Up and down is conventionally referred to as Y and left and right as X. The use of X and Y come from a convention that Descartes invented that has stuck, that known quantities should be given low letters in the alphabet, and unknown values letters towards the end of the alphabet, though I am yet to find anyone who can explain why the horizontal axis is X and the vertical Y. In Descartes' system, named after him as 'Cartesian coordinates', we can identify the exact position of any point on the paper (or whatever other two-dimensional object we're dealing with) by measuring off along the Y and X lines and producing a pair of numbers (see Figure 14.6).

So, for example, in the diagram below I have put three points on a sheet of paper. The one in the top right, labelled H is 1 unit along in the X direction and 1 unit in the Y direction, giving it the position $(1, 1)$. Point P, over to the left is 2 units up in the Y direction, but has gone below 0 on the X direction to -1, so its position is $(-1, 2)$. And so on. The reason mathematicians get so excited about Descartes' invention is that it enables them to represent the sort of lines and curves that we considered when looking at calculus as formulae. If, for example, we look back at

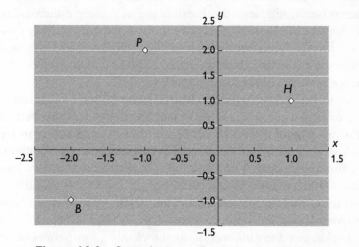

Figure 14.6 Cartesian two-dimensional coordinates.

the rocket's acceleration we saw on page 107, this can be described with the simple formula $x = y^2$. Just plug any value of y into the formula and you generate an x – plot the two with Cartesian coordinates and you get the curve. It makes it easy to swap between geometry and algebra, giving it a similar feel to one of the grand unification theories that physicists are always looking for – it's a way to pull together two major parts of mathematics in a unified whole.

However, in their excitement, mathematicians do tend to overlook the fact that long before Descartes was around – in fact for more than 2000 years – philosophers, scientists and geographers have been using this type of system to describe places in the world. Replace *P*, *H* and *B* on the diagram above with *The Plough*, *The Harrow* and *The Black Horse* and you have a map of the locations of three of our local pubs. Although the specific values of latitude and longitude as we now know them weren't formalized until relatively recently, the principle was already well known when Ptolemy drew up his world atlas in AD 150. The fact is, it's a very natural way of describing a location in two dimensions – to measure positions along two scales at right angles to each other. (Of course, the Earth is actually three dimensional, but because we stick to the surface in maps we can get away with two.) But this does not undermine the importance of Descartes' realization that this long-available concept could provide the bridge between algebra and geometry.

So, when Cantor moved up from the one-dimensional number line of 0 to 1, to a two-dimensional square, it was this sort of measure that he used. Just as one of the list of numbers in his table – say 0.52183828282... could represent a specific point on the line (in this case just to the right of the midway point), two such numbers, perhaps 0.3302829492... and 0.73472729019... could be used to identify a location within the square. Now Cantor wanted to apply the same type of approach as he had in working on the numbers in the 0 to 1 table, but he was faced with a more complex structure.

With the flash of originality that marks out genius, Cantor spotted an elegantly simple way of taking the approach he had used with one dimension and making it apply to two. He would transform the pairs of coordinates into a list of single numbers. All he had to do was alternate digits between the first and second coordinates that provided the location of a point on his square. If the numbers happened to be

$$0.3302829492\ldots \quad \text{and} \quad 0.7347272901\ldots$$

his new number would be 0.37330427822792499021 . . .

This is wonderfully clever because that single number contains all the information that was in the other two. So now he had a list of unique numbers, each between 0 and 1, mapping out every single point on the square. And this was no different to the list of numbers that mapped out the line. It was at exactly the same level of infinity. It had the same cardinality.

This left Cantor reeling. What he had proved was that one-dimensional space has exactly the same number of points as does two dimensions, or our familiar three-dimensional space – or for that matter as many dimensions as you like. You can perform exactly the same operation to transform the three coordinates that describe a cube – or as many dimensions as you fancy – into the single dimension of a line. Cantor wrote to his friend Richard Dedekind, a professor of mathematics in Brunswick, 'I see it but I do not believe it.'

Looking back on it now, it's hard to see why Cantor was so surprised, unless he had an expectation that there was some strong linkage between the levels of infinity and dimensions. After all, he already knew that aleph null stayed the same when it was squared. Galileo had effectively proved that $\aleph_0 \times \aleph_0 = \aleph_0$. What Cantor had now shown was that $\aleph_? \times \aleph_? = \aleph_?$. I have put that '?' in, because we're not sure what level of infinity the irrational numbers add up to. This seems remarkably logical, but to the mathematicians of the time, to whom the nature of dimensions was something set in stone, this was practically impossible. Perhaps we have a similar advantage to the White Queen in *Through the Looking Glass* when she responded to Alice's declaration that 'one *can't* believe impossible things':

> 'I daresay you haven't had much practice,' said the Queen. 'When I was your age, I always did it for half-an-hour a day. Why, sometimes I've believed as many as six impossible things before breakfast.' [96]

Exactly what this new bigger infinity, the infinity of the continuum, the range of numbers between 0 and 1, was in the grand scheme of $\aleph_?$ was something that Cantor was determined to establish. Was it the next infinity up the scale, or could there be something in between? Before hitting against the brick wall that this challenge was to prove, Cantor did manage to take one further step, thanks to his set theory.

One of the fascinating aspects of set theory that generates some of the more puzzling paradoxes associated with it is something called the power set. This frightening-sounding concept is, in fact, very simple. It says that for any set you choose, it is always possible to generate a bigger one using only the elements of that set. This is the set of all its subsets. Take a very simple set of the items in a (very small) fruit bowl – an apple, an orange and a banana. This is a set with cardinality 3 – there are three pieces of fruit. But the subsets are:

- Nothing – the empty set
- Apple
- Banana
- Orange
- Apple + banana
- Apple + orange
- Banana + orange
- Apple + banana + orange

... a total of eight different subsets. The cardinality of the subsets is 8 – much larger than the original 3. In fact, it turned out that for any set of things, the cardinality of the collection of subsets, the power set, was 2^c, where c is the cardinality of the set itself – in this particular case, $2^3 = 8$. This even works with the empty set, though in that case we are required first to work out what 2^0 is.

This is something that historically had troubled mathematicians, but was eventually settled on by looking at the way that the power, the small number written above, varies when you multiply numbers together. If we do the simple multiplication 4×8 we get 32. Now 4 is 2^2 and 8 is 2^3, while 32 is 2^5. So $2^2 \times 2^3 = 2^5$. Notice what happens to the powers. In doing the multiplication, the powers add up. It was Michael Stifel, the sixteenth-century German mathematician who was inclined to deny that irrational numbers exist, who first documented this.[97] To multiply 2^2 by 2^3 we *add* 2 and 3 to get 5. So what was 2^0? When you add this power to another you get no change. Multiplying by 2^0 produces no change. And there's only one number capable of that behaviour – the number 1. So 2^0, it was decided, was 1. (In fact, anything to the power 0 is 1.) The power set of the empty set has 2^0 or 1 element – the empty set itself.

These power sets were to prove very valuable in helping to understand the greater infinity of all the numbers, rational and irrational. As we have already seen, this greater infinity is often called the continuum, or c for short – as it represents the smooth, continuous sequence from 0 to 1 with no breaks.

Cantor discovered that the continuum itself is the power set of the integers. This made that greater infinity which might or might not be $\aleph_1$ to be $2^{\aleph_0}$. He had opened up one more fact about these special numbers that were infinity and more, the transfinite numbers. While adding or multiplying infinities didn't move up a level of complexity, multiplying a number by itself an infinite number of times did.

The argument as to why the continuum was the power set

of $\aleph_0$ is relatively simple to follow, but harder to prove in mathematical terms. Imagine every single number between 0 and 1, as we have before, in a huge table. But this time, instead of writing those numbers as we normally do on a counting system that goes from 0 to 9, let's use the binary system that all computers use internally. In binary, all numbers are represented using only 0's and 1's. The simple counting numbers become:

$$0, \ 1, \ 10, \ 11, \ 100, \ 101, \ 110, \ 111, \ldots$$

We can still represent all the numbers between 0 and 1 this way, although it takes a little thinking about. One-half, for instance is now 0.1 – you can tell this because by adding $0.1 + 0.1$, the binary arithmetic moves the 1 up one position to make 1.0. Each 1 we add to the number after the decimal point carves out another half of what's left. So 0.11 is made up of $\frac{1}{2}$ for the 0.1 and half of $\frac{1}{2}$ for 0.01, making $\frac{3}{4}$ – again you see if we add two lots of 0.11 together we get 1.1, or $1\frac{1}{2}$.

Once we've written all the numbers in our infinite table in this new form, we find each number has aleph-null digits – the number of positions in the infinitely long decimal fraction – each of which can have two values 0 or 1. Whenever there's a set of things that can have two values, the number of combinations is 2^n, where n is the number of things – the good old power set. So we have in this case $2^{\aleph_0}$ possible numbers.

To fit in with the sense of order that all of us feel, it would seem all too sensible that this next level of infinity, the infinity of the continuum, should be $\aleph_1$. Cantor could not prove this, but he believed it was so. After he had got so close, it might have seemed to Cantor that achieving the final goal of proving this continuum hypothesis would be relatively trivial. In fact it proved nightmarishly intractable. It was the challenge of dealing with the continuum hypothesis that would push Cantor over the edge into madness with increasing regularity. Or rather, it was the challenge of dealing with this problem when faced with increasingly virulent attacks from some of his colleagues.

15

MADNESS AND SANITY

Who shall tempt with wand'ring feet
The dark unbottom'd infinite abyss

John Milton, *Paradise Lost*, 1.404

LIKE ANY GREAT SHIFT in the way something is viewed there were many who clung on to the conventions of the time and refused to accept Cantor's novelties. Cantor may have been like Max Planck, father of quantum theory, who thought it was a neat, elegant theory, but clearly not believable, but others attacked Cantor's ideas just as 'classical' physicists resisted the quantum revolution (or for that matter those who put the Earth at the centre of the universe fought the Copernican model).

It didn't matter, as far as these conservative mathematicians were concerned, that Cantor had prefaced his 1895 article 'Beiträge zur Begründung der transfiniten Mengenlehre' (Contributions to the founding of the theory of transfinite numbers)[98] with a very telling quote from Isaac Newton. Cantor wrote *Hypotheses non fingo*, which Newton had written in his *Principia*. It means 'I frame no hypotheses' and the point Newton was trying to make was that all his deductions were made from reality – there was nothing based on arbitrary hypothesis in the manner of ancient Greek natural philosophy, where most theories were based on unsubstantiated thought rather than observation.

Why did Cantor take this comment from Newton? It would seem that he was saying that his work was not some arbitrary speculation, some top-of-the-head idea, but rather it was carefully worked out, based on observation and proof. It was not a view shared by his critics, however.

A typical example was Paul David Gustave du Bois-Reymond, a German mathematician, born in Berlin in 1831. Du Bois-Reymond worked almost exclusively in calculus, perhaps giving him a very particular view of infinity, of Aristotle's potential infinity, and he dismissed Cantor's findings with the derisive 'it appears repugnant to common sense'.

Another, and altogether more dangerous, opponent was Leopold Kronecker. This was a classic case of the mentor turning on his protégé. It seems a similar picture to the transformation in 1821 of the scientist Sir Humphry Davy from Michael Faraday's sponsor and friend to his worst critic. With Davy and Faraday there seems to have been a combination of insuperable class barrier – Faraday, a self-educated, working-class boy had acted as Davy's valet on a European tour – and an inability to accept that the amateur he had condescendingly allowed to play at being a scientist was showing signs of outclassing his mentor. For Kronecker it was not so much a matter of being from a different *social* class (though Kronecker was certainly the richer of the two) as being mentally outclassed. Cantor was thinking the unthinkable, treading on dangerous ground where Kronecker had no intention of following.

Leopold Kronecker was born in 1828 in Liegnitz in Prussia (now part of Poland). He showed mathematical ability from an early age, but after spending time at Berlin, Bonn and Breslau universities, and again at Berlin for his doctorate, he had to return home to deal with family business. He stayed in Liegnitz until 1855, by which time both independent of the family and independently wealthy he could concentrate on his real love, mathematics.

With no money problems, Kronecker did not particularly

want to be employed by a university. He returned to Berlin where he could meet with the other significant German mathematicians of the day without the irritation of dealing with the day-to-day problems of students. He had all the benefits of the university environment with none of the burdens. In 1861 the final piece was put in place in transforming him from a dilettante to a working mathematician. He was elected to the Berlin Academy, which meant despite his lack of formal links with the university he could lecture there on subjects that interested him. It was in this position that his lectures were attended by Cantor, and that later he would give Cantor advice on getting started in his new career at Halle. Kronecker had become well known and respected, with strong connections throughout the academic hierarchy.

This was the position of power Kronecker held when he first came across Cantor's new theories. Whenever something new sweeps a discipline – whether it's quantum physics displacing classical physics or the latest style of pop music arriving on the scene, there will always be those who long for things to stay the way they were. Kronecker was worse than this. In physics terms he would have been saying things started going wrong with Aristotle, while musically, he would have been suspicious of any song written later than William Byrd (and even Byrd he would have harboured some doubts about).

It's not just that Kronecker had problems with Cantor's admittedly challenging ideas of the nature of infinity. He was a nineteenth-century mathematician who only really accepted the existence of integers (and hence rational numbers). Despite being a superb theoretician, he was uncomfortable with much of the work that had been done since the Pythagoreans.

It might seem that Kronecker was simply living in the past. Here was a man who did not accept the validity of $\sqrt{2}$ or π or imaginary numbers, all of which had been standard contents of the mathematician's toolbag for hundreds of years. Yet, though this approach might appear similar to that of an ostrich hiding its head in the sand, Kronecker would probably argue that all he was doing was applying a scientific method to mathematics.

Many mathematicians are happy to work their numerical magic in total isolation from reality – in fact to these pure mathematicians, to do anything else misses the whole point of mathematics. It's a superb game, an intellectual challenge, not something that should be looking for any application in the real world. If, however, mathematicians take a more scientific, rather than game-playing approach, they will want not just proof but demonstration. They will want to be able to see the reality of their mathematics, and once you go beyond the integers, there are some real problems.

We can demonstrate the integers with, for example, a basket of fruit, counting out individual pieces. Similarly, rational numbers that consist of one number divided up by another can be demonstrated easily with that same fruit basket. If you take an entity called 'a dozen apples' you can exactly demonstrate what half of it, or three-quarters of it is, with no approximation. Negative numbers become a little tougher. I can't show you -3 oranges. But it is relatively easy to convey the concept of a negative number – for example, if I have a row of ten oranges and take three out, the gaps in the row could be considered to be negative oranges. Equally, we can deal easily with integers and rational numbers on a calculator.

However, as soon as we reach the irrational numbers, as we have already seen, there are problems. In the real world we have to resort to approximation. You might have a π button on a calculator, but it will only ever produce a few decimal places, then stop. You might measure the ratio of a circle's circumference to its diameter, but once again it will result in a crude approximation. Measuring the diagonal of a 1 metre square will, again, approximate to $\sqrt{2}$, but you will never be able to produce that true number in all its glory. As for ideas like $\sqrt{-1}$, even though this concept is very valuable in real-world calculations, there is no physical entity that can be used to directly demonstrate it.

Kronecker's stance, then, while not one shared by many mathematicians, or for that matter many scientists or engineers, is by no means illogical. And given that position, it is not surprising that he found Cantor's elegant manipulations of infinity to be distasteful in the extreme. Cantor had built a whole structure on the irrational numbers. As far as Kronecker was concerned, that structure had no foundation at all. It was capable of being blown away by the least impact.

This explains why Kronecker was to turn against the theories of his one time protégé, though could not really explain the emotional force with which he took up his arguments. When earlier I likened this breakdown of an intellectual relationship to the way Michael Faraday's mentor, Sir Humphry Davy, turned against him, we have two situations with very different triggers.

Davy seems to have had a very patronizing attitude towards Faraday, never expecting any great originality from the man who had been his lab assistant and (briefly) valet. When he rounded on Faraday, it was because he thought that Faraday had been guilty of plagiarism, stealing an idea from Davy's friend (and social equal), William Wollaston. Although all the evidence pointed away from this being the case, Davy could never go back on his initial negative judgement, even going so far as to vote against Faraday when his discovery of the phenomena that made the electric motor possible led to his otherwise unanimous election to the prestigious Royal Society.

In Kronecker's case there was no suggestion of malpractice by Cantor. Kronecker simply could not accept the theories of his former student, and found them so distressing, so far removed from the reality that could be proved with the senses, that he began an all-out campaign to discredit Cantor and his theories. Arguably Kronecker was a fanatic. He realized that he held a minority view, and so was wounded by anything that brought unnatural numbers into mathematics. Seeing that Cantor was not only coming up with these theories but teaching them, he called him a corruptor of youth, someone who took away the youthful innocence of a dependence on reality, leading his students into a dangerous fantasy of horribly unnatural numbers. It would not be stretching things too far to say that Kronecker regarded Cantor as a mathematical pornographer.

To be fair to Kronecker, Cantor also seems to have been highly emotional about his theories. Harking back to St Augustine, he was prepared to equate the ultimate infinity with God, seeing a stately progression from the real numbers to the first level of infinity, then to the second, the infinity of the whole set of numbers between 0 and 1, and on through unknown further levels to the insuperable infinity of infinities, God himself. By working on the mathematics of infinity, Cantor saw himself as expanding our knowledge of both numbers and the divine.

Two individuals with such diverse viewpoints are rarely capable of compromise. You can bring them together at a table for discussions, but you are liable to end up with the initial positions becoming more and more entrenched. In a political dispute, settling this type of conflict will usually require a third party's intervention. In the case of Cantor and Kronecker there seems not to have been a suitably strong individual who could bring the two sides together. The two camps were painfully divided. Cantor did make one attempt at reconciliation, meeting up with Kronecker in informal surroundings to try to iron out their differences, but without an impartial referee, the result was predictable. There was no movement on either side.

It might seem, with the softening distance of a hundred years in time, and the now universal acceptance of much of Cantor's work that this would have been a very one-sided argument. After all, while Kronecker wasn't alone in his views, they could hardly be regarded as the accepted wisdom of the time. But the academic world has always been susceptible to political manoeuvring, and never more so than in late-nineteenth-century mathematics.

Cantor had undisputable mathematical logic on his side, but Kronecker had the political edge. He proceeded to make the most of his advantage. Kronecker was a significant power in

Berlin academic life, while Cantor taught at the second-rate university at Halle. Kronecker had plenty of friends in the establishment. Cantor, who was not in any sense a big name, seemed to go out his way to irritate the academic powers, making it clear that he thought himself better than his colleagues, and more worthy of a post in Berlin than many of those who held them at the time.

Kronecker couldn't bring academic circles round to his own way of thinking, but he could obstruct Cantor's career and his chance to get his papers published, and this he proceeded to do with remarkable ease. Kronecker was able to do this because of the way that scientific and mathematical and scientific papers reach the world – a process known as peer review. This system is designed to ensure that only papers that have a good chance of being valid get published, but it is a system that depends on the honesty and straightforwardness of those operating the system. While usually very effective, it has a worryingly ready potential for corruption. Interestingly, more than a hundred years after Cantor had problems publishing his papers, the whole system would be placed under intense scrutiny as the scientific world wondered whether or not this was the best way forward.

Read an article in a normal magazine or a newspaper and you are seeing a product that has probably only passed a relatively small amount of scrutiny on its way to the printed page. The magazine editor might have picked up on a point or two. The sub-editor will have scanned the piece for typographical errors, and perhaps trimmed it to fit the space available. But it will have been down to the author to determine whether or not the facts presented in the article are true.

This level of checking is fine for most content (and all that is practical in the sort of timescales many publications operate on), but isn't appropriate for the scientific world. Instead, the paper is sent to a number of expert reviewers, who check it and submit comments that may result in parts of the paper being rewritten, or the paper being withdrawn altogether. This peer-review process is anonymous – you don't know who your reviewers are – and is intensely time-consuming.

The peer-review process has recently come under question. In two cases in the last few years, papers were published despite unfavourable peer review. In 1999, the British medical journal *The Lancet* published a paper on the effect of genetically modified potatoes on rats that some of the reviewers urged the journal not to publish. In 2002, the top UK-based science journal *Nature* published a paper, swiftly followed by another paper with a totally opposing view alongside a third paper from the original paper's authors backing up their original findings. The peer reviewers disagreed over whether the first paper should ever have been published, so the journal decided to publish everything, while admitting that perhaps they it had made a mistake:

> In light of these discussions and the diverse advice received, *Nature* has concluded that the evidence available is not sufficient to justify the publication of the original paper.[99]

It went on, though, to justify the further publications by saying:

> As the authors nevertheless wish to stand by the available evidence of their conclusions, we feel it best simply to make these circumstances clear, to publish the criticisms, the authors' response and new data, and to allow our readers to judge the science for themselves.[100]

This 'make up your own minds' attitude was also the reasoning behind the decision of *The Lancet's* editor to publish. If their reasoning is true, it suggests that peer review is always in danger of suppressing new ideas that don't fit with the current accepted picture of the world – that, worthy though the process is, it naturally operates against the development of truly new ideas. Of course, simply taking peer review away is no answer. As the UK magazine *New Scientist* commented at the time:

> By leaving readers to make up their own minds about research a reviewer rejected, they seem to be failing in their duty—but show the

confusion caused when peer review goes wrong. Nobody knows what to believe.[101]

The *New Scientist* editorial went on to suggest that

on some important, contentious issues, science relies too heavily on peer-reviewed journals. To make sense of these cases we need a different system altogether, and you don't have to look far for a model. Bodies such as Britain's Royal Society and the US National Academy of Sciences already initiate investigations into broad areas of science. It wouldn't be much of a stretch for them to form scientific 'hit squads' to report quickly on narrow but important issues.[102]

Of course, even the *New Scientist* editor's suggestion is open to some doubts. After all it was one of the bodies named – the Royal Society – that set up the less than effective 'hit squad' to address the priority debate over the invention of calculus in 1712 and managed to totally rewrite history. It might be tempting fate to suggest that such a thing could never happen again.

Back in Cantor's time there was no question of changing the review process, but Kronecker was capable of infiltrating it, of either becoming a reviewer himself, or persuading reviewers or the journal in question that there were so many problems with Cantor's papers that it would be dangerous to publish them. When Cantor submitted his paper on the way that different dimensional bodies – a line, a square, a cube, and so on – were all identical in the number of points they contained, he expected the journal, a respected German mathematical publication usually referred to as 'Crelle's journal' to get his paper into print reasonably quickly. But the months went by and nothing seemed to be happening.

At the time, the journal was a well-established part of the German mathematical world. It was founded by August Leopold Crelle, a German civil servant with a self-taught expertise in mathematics, in 1826, and was properly called *Journal für die reine und angewandte Mathematik*. Crelle himself continued to act as chief editor for the first 52 volumes, and prided himself on his ability to spot fresh new talent and bright ideas, but he had been dead for 22 years by the time Cantor submitted his paper in 1877, and his successors at the time were less open-minded than Crelle himself had been.

One of the editors finally confided in Cantor that the delays were caused by a constant barrage of negative comment from Kronecker, assuring the editorial staff that publishing such a paper was a waste of space, that it was a non-paper, meaningless and useless. Crelle's journal did eventually publish, but Cantor would never make use of them again. Cantor had just about won the first skirmish, but Kronecker was by no means finished with his one time student. Over the next few years he kept up a steady stream of bad publicity, seemingly well aware of the fact that repeated allegations of a negative nature are usually enough to weaken a public image.

Cantor knew that Kronecker's criticisms had become more and more personal, attacking the man as much as the theories, but he was unable from his backwater location in Halle to have much impact on the Berlin gossip circle. By 1883, Cantor decided to take the battle more directly to Kronecker. He applied for a professorship in Berlin. This was, to be fair, a position that Cantor desperately wanted – but he knew very well that Kronecker would block him, and (at the risk of getting into tangled logic) Cantor knew that Kronecker knew that Cantor knew he would block him – and his Berlin-based rival could only see Cantor's application as a personal attack. As Cantor was to write to the Swedish mathematician Gösta Mittag-Leffler:

> I know precisely the immediate effect this would have, that in fact Kronecker would flare up as if stung by a scorpion, and with his reserve troops would strike up such a howl that Berlin would think it had been transported to the sandy deserts of Africa with its lions, tigers and hyenas. It seems that I have actually achieved this goal![103]

Unfortunately, Cantor's attempt to get at Kronecker backfired.

His rival found a way of hitting back that was more effective than Cantor's own ploy.

At the time, Cantor was publishing papers in the journal run by his friend Mittag-Leffler, called *Acta Mathematica*. This was in some ways a surprising friendship. Magnus Gösta Mittag-Leffler spent much of his life in the city of his birth, Stockholm in Sweden, but he was to discover the German mathematical scene as a result of a requirement of his course at the University of Uppsala. To gain his degree, Mittag-Leffler had to spend three years abroad, which he split between Paris and Berlin. He met Cantor very infrequently, but they kept up a long correspondence – Mittag-Leffler seemed to be one of the few people Cantor fully trusted to be able to share his ideas with.

When he discovered Cantor was using Mittag-Leffler's journal, Kronecker showed the depth of his inclination to psychological intrigue by offering to publish in *Acta*. The thought horrified Cantor. Kronecker had managed to destroy his image in Berlin – if this well-established, powerful mathematician began to publish in the only journal where Cantor felt safe it would mean that there was nowhere to hide, nowhere that he could publish without sniping and criticism from a man for whom he now felt nothing but hate.

As Kronecker seems to have anticipated, Cantor blundered in and tried to be heavy handed. Rather than simply voicing a concern to Mittag-Leffler, Cantor played the prima donna, threatening to take his work elsewhere. Not surprisingly this made relations between the two strained. And Kronecker's paper? It never seems to have appeared. Chances are, it never existed. The non-existent paper was simply another virtual weapon in a battle to destroy the spirit. And it succeeded. Within a few months, Cantor had undergone his first nervous breakdown.

The resistance to Cantor's theories was strong initially, but just as quantum theory was to become the bread and butter of the physics world, set theory and transfinite numbers would be accepted into the core of mathematical knowledge as standards of the field. There may have been some wishful thinking in another of the quotes Cantor put at the start of his article 'Beiträge zur Begründung der transfiniten Mengenlehre' (Contributions to the founding of the theory of transfinite numbers),[104] yet in the long run his use of Seneca's *Veniet tempus, quo ista quae nunc latent, in lucem dies extrahat et longioris aevi diligentia* (The time will come in which the diligence of a longer age draws into daylight those things that are now concealed) turned out to be an accurate prediction.

That time, though, did not come soon enough to rescue Cantor from his mental distress. Year after year he plunged between depression and elation as he first believed he had disproved his theory, and then believed he had proved it – only to discover that his proof was irredeemably flawed. Cantor's periods of nervous breakdown grew longer and longer until his death in 1918.

Cantor's new mathematics of infinity might have irritated mathematicians at the time, but as is often the case with a breakthrough that is initially questioned, it was not long before the once crazy seeming theories had been absorbed into the mainstream of mathematics. By the mid-1920s, German mathematician Hermann Weyl was able to call mathematics 'the science of the infinite', emphasizing the degree to which set theory with its inevitable contemplation of the infinite was at the heart of mathematics. These days a university text on set theory doesn't even bother to mention that there was ever any doubt about Cantor's work.

One man carried Cantor's work on infinity further to discover a totally unexpected and frustratingly terminal result. Cantor was to be followed into the infinite and into madness by another mathematician who made an in-depth study of the continuum hypothesis – of whether the infinity of all the numbers between 0 and 1 was indeed $\aleph_1$ or was instead another aleph. His name was Kurt Gödel.

The Gödel family were technically Czech, and Kurt was born in Brno in 1906, but his family was of German background, speaking German at home and bringing up Kurt and his older brother Rudolf very much as if they were living in a German or Austrian home. It seemed entirely sensible, then, that Kurt, who was already showing signs of great skill in mathematics, should attend university in Vienna. Although the mathematics department there was not so famed as that of Berlin, Vienna was relatively close to Brno, and so more practical to keep up family ties. What's more, Rudolf was already there, making it the ideal choice for his younger brother.

Vienna continued to prove an attractive location, and Gödel stayed on for his doctorate and subsequent research. Unlike Cantor, Gödel was no social recluse – quite the reverse. He seemed capable of partying all night and still coming up with quite remarkable new ideas. At one of the night clubs he attended, Gödel met a dancer, Adele Porkert, older, more sophisticated – exactly the kind of woman his mother probably warned him against. Nonetheless, Adele was to become Mrs Gödel. With Adele at his side, the partying went on, as did Gödel's rapidly maturing ability to challenge the accepted norms of mathematics.

Gödel was to come up with one of the most shocking proofs in all of mathematics, one that would have a direct impact not only on set theory, but on every part of mathematics. His bewildering masterpiece is called the *incompleteness theorem*. In essence it states that in any system of mathematics above a certain level of complexity there will be some problems that will be inherently insoluble. (A system of mathematics is the basic set of rules – the axioms – from which mathematicians then derive proofs of what is and isn't true.)

According to Gödel, no matter how much effort is put into some problems, they can't be cracked. It's a bit like trying to throw a tennis ball by hand fast enough to get it to escape from gravity and shoot off into space. Unless you move into a new system, where instead of throwing it by hand you use some artificial means of propelling it, you will never be able to throw it fast enough to escape the gravitational field. It doesn't matter what throwing technique you try, how much you practice, how much you build your muscles, you won't be able to do it. The same goes for some problems in mathematics: given a particular mathematical system, it doesn't matter how you approach the problem, you can't solve it.

A crude approximation to Gödel's theorem is to imagine you had to deal with the statement 'This system of mathematics can't prove this statement is true'. Ask yourself, whether or not this statement would be true. If the system proves the statement, then it can't prove it. If the system can't prove the statement, clearly it still can't prove it. Whatever happens, this is a statement that the system can't prove – so we have just established that, whatever you system of mathematics there are statements it can't prove.

That might seem a bit like playing with words. It feels rather like asking whether or not the statement 'This is a lie' is true. But Gödel did not rely simply on rather tenuous use of language – he was able to find a way to define an actual mathematical statement that cannot be proved in any system of mathematics.

As Gödel began to work more and more on infinity and the continuum hypothesis, like Cantor before him he began to show signs of nervous breakdown. In Gödel's case, there was a strong element of paranoia. He believed that others were trying to drive him out, or even to murder him. Gödel did not have an equivalent of Kronecker and Cantor's other opponents, but it seemed that once more the contemplation of true infinity was endangering a great man's sanity. And for Gödel there was another factor involved.

In the mid-1930s, as the rise of Nazism made Vienna an increasingly dangerous place, Gödel was invited to join the Institute for Advanced Study at Princeton in the USA, which already had Albert Einstein on its books. Gödel was not a Jew, but many of his colleagues were, and he was attacked in the street on the suspicion of being Jewish. It would seem that an escape to Princeton would have seemed very attractive, but Gödel did not last long there, returning home after only six months.

In the increasingly unstable atmosphere of 1930s Austria, Gödel was to take a half-way step to Cantor's fervent belief that the infinity of all numbers between 0 and 1, the infinity of the continuum, was $\aleph_1$. Despite his worsening mental illness, he was able to reduce the barriers to the continuum hypothesis. This discovery was made in 1937 (it seems, according to a short, rather obscure, note, on the nights of 14 and 15 June), but was not written up until the 1940s. Gödel proved not that the continuum hypothesis was true, but rather that it wasn't inconsistent with the axioms of set theory. In other words, he could not say for definite that it was the case, but if you assumed it was true it didn't upset any of the workings of set theory. This might count as the basis for an acceptable theory in physics or the other sciences, but for mathematics is simply not enough – it's no proof at all.

A combination of obsession with his work and mental illness seemed to shield Gödel from the horrors of the Nazi regime. It was only in 1939, when the authorities informed him that he had been declared fit for military service that Gödel seems to have realised the situation he was in, and just managed to scrape through with his wife to leave the country for America before all possibilities of travel were closed down. It was already too late to take the western route. Instead they risked the Trans-Siberian Railway, travelling on to Japan and from there by boat to San Francisco.

Once in America, Gödel's health became worse. His paranoia deepened. Though untangling the continuum hypothesis remained his theoretical goal he was distracted more and more by obscure passions. Could he prove God's existence mathematically? Was the mathematician Leibniz credited with less discoveries than he had actually made? While on a holiday, the distracted Gödel was nearly arrested as a spy as he paced along the seafront, muttering in German to himself. The locals thought he was waiting to contact a U-boat.

Gödel was never to get any further on the problems of infinity. Although he lived on to 1978, his last paper was published in 1958, when he was only 52. His health in later years was poor. His brother, Rudolf, who had become a doctor, put this down to Gödel's belief in his own medical knowledge:

> My brother had a very individual and fixed opinion about everything and could hardly be convinced otherwise. Unfortunately he believed all his life that he was always right not only in mathematics but also in medicine, so he was a very difficult patient for doctors. After severe bleeding from a duodenal ulcer . . . for the rest of his life he kept to an extremely strict (overstrict?) diet which caused him slowly to lose weight.

But Gödel's strict control of his diet seems to have been mostly due to his conviction that others were out to poison him, forcing him to eat only food that he could be sure had not been tampered with. Was Gödel really driven mad by contemplation of infinity? Like Cantor before him, he had another contributing factor, in his case the growth of Nazism in Austria. Yet, also as with Cantor, the external factor seems not to have been enough on its own. Gödel mostly seemed to have little interest in what was going on around him at a political level. Once more it does seem that the attempt to see infinity clearly destroyed a great mind.

It was another, younger mathematician, Paul Cohen, based at Stanford, who would take the final step in showing whether or not the continuum hypothesis – that the infinity of the real numbers was indeed $\aleph_1$ – was true. But with the frustrating intangibility that seems to haunt the theory of the infinite, Cohen would neither prove the continuum hypothesis to be true, nor would he prove it false.

Instead, Cohen showed that the continuum hypothesis is independent of the axioms of set theory. Where Gödel had shown that there was no contradiction between the continuum hypothesis and set theory, Cohen managed to prove that set theory would still work if the continuum hypothesis was untrue. In other words it will never be possible to prove the hypothesis either way, since no inconsistencies arise with the set theory axioms whether the continuum hypothesis is true or false. It is said to be 'undecidable' within standard set theory. In principle there could be a new set of axioms – a new way of defining how sets work – that would enable the hypothesis to be proved or disproved, but with the standard approach to sets, the one still used today (called the Zermelo–Fraenkel system after the mathematicians who formalized it), there could never be a useful outcome.

We have already met these axioms, but not in any detail – the concept, though, is quite simple. Axioms are mathematics' givens – the assumptions that are made to begin with, the building blocks from which theorems are built. Mathematicians tend to give these axioms rather grandiose names, even when they are reasonably simple. The Zermelo–Fraenkel axioms for sets are:

1. **The axiom of existence – there exists at least one set.** Remember how the cardinal numbers are built from the empty set, the set with nothing in. But you do need something to start with.

2. **The axiom of extension – two sets are equal if and only if they have the same elements.** Seemingly obvious, this is the sort of 'common-sense' suggestion that has to be formally stated as an axiom for mathematics to be properly consistent.

3. **The axiom of specification – for every set and every condition, there corresponds a set whose elements are exactly the same as those elements of the original set for which the condition is true.** That is, however you choose some of the elements from a set, those elements will themselves form a set. So, for instance, if my condition was 'the number is even' the application of that to the set of all integers will produce the even integers – itself a set.

4. **The axiom of pairing – for any two sets there exists a set to which they both belong.** So effectively you can make a set out of two other sets.

5. **The axiom of unions – for every collection of sets there exists a set that contains all the elements that belong to at least one of the sets in the collection.**

6. **The axiom of powers – for each set there exists a collection of sets that contains amongst its elements all the subsets of the given set.** Remember the power set of apples, oranges and bananas.

7. **The axiom of infinity – there exists a set containing the empty set and the successor of each of its elements.**

8. **The axiom of choice – for every set we can provide a mechanism for choosing one member of any non-empty subset of the set.**

It is the last axiom, the 'axiom of choice' that has caused the most problems for those attempting to prove the continuum hypothesis. Zermelo assumed it was possible to make the choice of an element, even from an infinite subset – but gave no guidance as to how that choice was going to be made. Although we can easily imagine just picking out one element, there is a human interaction there that isn't present in the mathematics. How is the choice to be made? What are the rules? This is something that makes mathematicians uneasy, and there remains doubt to this day that this axiom is really an appropriate foundation on which to base mathematics. Cohen showed that this axiom, along with the continuum hypothesis, was totally separate from the rest of the axioms.

It might seem strange to say that one of the axioms in a set is totally unrelated to the others – but it is rather like looking at two unrelated aspects of a physical object. If we take a ball it has mass and spatial dimensions. The two are intimately involved in the way the ball acts, and both can be changed by various actions (for example, as we saw when looking at Einstein's special relativity, if you accelerate the ball to near the speed of light, its mass will get larger and its spatial dimensions will get smaller). But there is no way of making a direct linkage from one to the other, lacking any other information. I can't tell you how big a ball is, given its mass, nor what its mass is given its dimensions. Similarly, the axiom of choice and the continuum hypothesis are all part of the same mathematical structure, but there is no relationship between them and the rest of the axioms: they are entirely independent.

As long as we continue to use these axioms as the 'givens', the things we take for granted about sets, then there can be no progress made on the continuum hypothesis. Ever. The French-born mathematician André Weil, who did most of his work in the US after moving there during the Second World War, summed up the frustration that inevitably arises from the contradictions of modern mathematics with a neat phrase:

> God exists since mathematics is consistent, and the devil exists since we cannot prove it.[105]

16

INFINITESIMALLY SMALL

It has long been an axiom of mine that little things are infinitely the most important.

Arthur Conan Doyle, *The Adventures of Sherlock Holmes*
(A Case of Identity)

WHETHER OR NOT YOU ACCEPT that there is such a thing as infinity for real, it's a grand, sweeping concept. At first sight, it's hard to get as excited about the opposite end of the number size spectrum – the incredibly small. The standard answer to 'what do you get if you divide 1 by infinity' is 0, yet shouldn't we instead be dealing with a submicroscopic number, the minuscule equivalent of infinity itself? Something that isn't 0, but is infinitesimal?

As we have seen, the infinitesimal rears its head everywhere in calculus. Just as infinity was made virtual to facilitate practical uses, so was the infinitesimal. Leibniz emphasized that what was meant by infinitely small quantities was not this in truth, but rather something that was 'indefinitely small', which he went on to clarify as being 'as small as you please'... while still remaining an actual value.[106] The philosopher Bishop George Berkeley, you will remember, refers to infinitesimals as 'ghosts of departed quantities'. At first sight it seems easy enough to prove that infinitesimals just don't exist.

Imagine we had the smallest number there could be – call it

ghost. Now divide it by 2. We've just produced a number that is bigger than zero (because two of them added together make *ghost*), yet is smaller than *ghost*. So *ghost* wasn't the smallest number there could be.

Yet this 'proof' depends on a critical assumption – that infinitesimals obey ordinary arithmetic, rather than the arithmetic of infinities. What would happen, though, if they were a very different kind of number, just as $\aleph_0$ is a different kind of number? We know that $\aleph_0 \times 2 = \aleph_0$, so why should it not be the case that $ghost/2 = ghost$?

All this concern with making infinitesimals unnecessary for calculus resulted in the whole business being much more complicated than it needed to be – yet for many years mathematicians were convinced it was the only route to sanity. It was only in the 1960s that the key to making infinitesimals respectable and usable was spotted. To understand the idea, it's necessary to take a trip back to the unreal. Remember imaginary numbers, based on the use of i, the square root of -1. They combine with real numbers to form complex numbers, something like $3 + 7i$, where '3' is a perfectly ordinary number, but the $7i$ part depends on the intangible concept of $\sqrt{-1}$. To anyone short of a Leopold Kronecker, this is a valuable and useful part of the mathematical toolkit, even though it isn't possible to draw a line that is $\sqrt{-1}$ long or pick up $\sqrt{-1}$ oranges.

So, for hundreds of years, everyone had been quite happy about having a special sort of number, part of which was in the 'real' world and part of which depended on a physically inconceivable construct, the square root of -1. Abraham Robinson had the flash of inspiration that infinitesimals could be made acceptable in just the same way.

Abraham Robinson (originally Robinsohn, but changed to Robinson in 1940) was one of the many Jewish scientists and mathematicians who decided it would be wise to leave Nazi Germany in the 1930s. He was born in Waldenburg (now called Walbrzych and part of Poland) in October 1918 into an intellectual family. His father, a philosopher and writer, died soon after Abraham was born, leaving him and his older brother to be brought up by his schoolteacher mother. The Robinsohns had contacts in Palestine, and early in the Nazi discrimination against Jews, in 1933, they moved to Jerusalem.

By now Robinsohn was showing a strong aptitude for mathematics, and by the end of his degree course at the Hebrew University in Jerusalem he was able to win a scholarship to the Sorbonne in Paris. The timing was terrible. He arrived in 1939, and within months was fleeing the German invasion, heading south to Bordeaux, where he managed to find a small boat to take him to England. It was on his arrival that he changed his surname to Robinson. He was to stay in the United Kingdom, first at the Royal Aircraft Establishment at Farnborough and then at London University until 1951, when he transferred to Toronto, and then back to Jerusalem in 1957 to take over the mathematics chair.

It was here, in 1961, and over the next few years after his move to UCLA in 1962 that he made the breakthrough that was to make infinitesimals an accepted part of the mathematical toolkit.

In essence, the original idea was recognizing that infinitesimals were outside the standard number system – that just like imaginary numbers they operated in something akin to a different dimension. It was no more meaningful to say that an infinitesimal was between the smallest 'real' number and zero than to try to fit i, the square root of -1, somewhere on a number line between -2 and $+2$. Infinitesimals could be treated as a new class of numbers, with their own mathematical operations. And, just like the complex number that has real and imaginary components, there could be a complex number with standard (i.e. not infinitesimal) and non-standard components. Newton's $2x + o$ is a classic example of this – the $2x$ being the standard component and the ghostly o the non-standard.

It wasn't, of course, just a matter of saying 'let's call infinitesimals a different kind of number'. Robinson used a technique called model theory to show that the same approach that gives a formal structure to define how real numbers combine together through the familiar processes of arithmetic, and how they follow in order, can be extended, stretched to include the infinitely large *and* the infinitely small. These infinitesimals might not be the same as real numbers, but they had similar properties and could be operated on in similar ways – just as was the case with imaginary numbers.

Robinson's approach became known as non-standard analysis. It provides a means to accept the intuitively obvious possibilities – like Nicholas of Cusa's orange segments of a circle, or Newton's fluxions – but to give them the sort of rigorous treatment that mathematicians could accept. No longer was it necessary to worry about dividing by zero as the infinitesimal faded to nothing, you were dealing with acceptable, if non-standard, mathematical quantities.

Infinitesimals aren't just a way of keeping mathematicians happy with something that's obvious to everyone else. Just like imaginary numbers, they have proved effective tools in mathematics and physics to deal with problems that simply couldn't be handled any other way. It would be easy to think of infinitesimals as a kludge – taking a dubious old theory and glossing over the problems to make it work – but it's not like that. Non-standard analysis is a rigorous mathematical discipline, not a version of Newton's ghosts with a new image.

The situation has many parallels with another of Newton's ideas that was not strongly founded on fact, but has surprising similarities with a modern theory. Newton was convinced that light was made up of tiny particles – corpuscles, he called them – flying through space. When the wave theories of light were proved, Newton's theory was reluctantly dismissed. But in the twentieth century we learned that light is as much like a particle as it is like a wave. It wasn't that Newton was vindicated, but the reality, based on solid experiment and theory, had a similarity to Newton's ideas. The same is true of the difference between present-day infinitesimals and those once used with such cavalier enthusiasm in the seventeenth century.

To be precise, infinitesimals are defined as sitting on a number line (mathematicians call the special number line including infinites and infinitesimals as well as the conventional real numbers the *hyperreal* number line) and being bigger than $-a$ and smaller than a for all values of a. They hover between the smallest negative and the smallest positive that can exist. It would seem that this makes 0 an infinitesimal, and for once the mathematics of infinity does obey common sense – it is. Zero is the only real infinitesimal, the only one we can work with in normal mathematics, but alone it isn't really of any use – non-standard analysis needs us to bring in a whole cloud of other non-real numbers that sit in the gap between $-a$ and a.

Even so, most of us who were taught calculus will still raise an eyebrow, surprised by the apparent validity of infinitesimals. Non-standard analysis is something that many non-mathematicians still don't realize exists. But the fact is that for mathematicians these are well-established techniques, dating back decades, with sober textbooks to support them.

In areas where they have found application, infinitesimals and the different viewpoint of non-standard analysis have enabled previously intractable problems to be worked on afresh. It seems particularly useful in areas of physics dealing with small particles. A good example is Brownian motion, the theory that describes the way small particles like dust are buffeted around in a random way by collision with molecules.

The name dates back to the 1820s, when the English botanist James Brown noticed that pollen grains in a drop of water, viewed under a microscope, jumped and danced around in an unpredictable fashion. This was originally assumed to be some characteristic of the life in the pollen, but it was soon found that ancient pollen with no possibility of life remaining, behaved exactly the same way. It was in 1877 that Desaulx proposed the correct reason – that it was the natural thermal motion of the molecules in the liquid that was causing them to collide with the much bigger pollen grains and jolt them into motion – and not until 1905 that Einstein provided a mathematical description of what was happening.

The Brownian motion model has been applied to everything from stock-market movements to machine station queuing in manufacturing, but despite Einstein's involvement, it proved very difficult to model. The approach that would finally crack it was non-standard analysis and its infinitesimals. Robert Anderson, now professor economics and mathematics at the University of California, Berkeley, was a Ph.D. student at Yale in 1976 when he devised the approach. Rather in the way biologists split a microscope slide into a tiny grid to count populations of tiny organisms, Anderson split space on a computer into a three-dimensional grid.

Each cube of space on the grid could contain a molecule, and with each measure of time that passed the chunks of space were moved around like a huge, independently mobile sliding puzzle. By then making the chunks infinitesimally small, and allowing movements to take place in infinitesimally tight slivers of time it was possible to transform a jerky, step-by-step model into one that accurately reflected real life. Without non-standard analysis this approach had proved impossible – now it meant that Brownian motion could be properly modelled.

An approach with some similarities is being taken by Jean-Pierre Reveillés of the Louis Pasteur University in Strasbourg. Here an infinitesimal three-dimensional grid is being used to manipulate 3D images that are to be represented on a computer screen. The great advantage of using infinitesimals is that, unlike real pixels, there is no problem when rotating the image of working out how all the points on the image map on to your pixel array. There aren't going to be any sticky-out bits that don't quite fit, as you always tend to get with pixels. All the manipulation can be done this way first, and only at the last moment, when translating back to the physical screen, do you have to perform the heavy-duty calculations to match up the image to the pixels, vastly reducing the processing power required to manipulate complex images.

When talking about non-standard analysis, it is usually the infinitesimals that get stressed. These, after all, not only support the intuitive rightness of the slightly *risqué* results of calculus, but also provide openings for whole new solutions. However, Robinson's theory did also encompass the infinitely large as well as the infinitely small. It would be wonderful to think that Cantor's transfinite numbers were now brought, in some way, back into the 'normal' family of numbers.

Unfortunately this isn't the case. Cantor's infinite numbers can't be squeezed into a consistent number line with real numbers in the way Robinson's can. They remain counter-intuitive. The two systems are inherently incompatible. So does this mean that one is wrong and the other right? Not at all. Perhaps the best picture of what is happening here is to look at what happens when we capture a three-dimensional object onto the two-dimensional space of a piece of paper.

Take a railroad train. Photograph it sideways on and you will get a very long, thin picture. If that (and hundreds of other pictures of trains from the side) were our only evidence of what trains were like we could put together a description of trains that included 'they are much, much wider than they are high'. But then we come across a picture of a train taken from the front. It radically breaks the mould. It is actually higher than it is wide. Either this is a different thing altogether, or one of our descriptions of a train in wrong.

In reality, both are correct for the particular way we look at them. The same appears to be true for infinity. Whenever we deal with infinite mathematics in 'normal' number space we have to submit to a similar reduction in information to the move from three dimensional space to the two-dimensional picture. When we look at infinity from one 'direction' we get Cantor's alephs and omegas. From another we see Robinson's non-standard analysis.

As K. D. Stroyan and W. A. J. Luxemburg say in their textbook *Introduction to the Theory of Infinitesimals*:

> We see no conflict; in fact we believe important mathematics will continue to develop out of the interaction of the two and that neither theory is 'real' since they do not submit to experiments.[107]

We are no more likely to get a complete picture of infinity than a two-dimensional creature could ever understand the real train. If anyone could, they would have to move into the absolute understanding that is ascribed to God by some of the philosophers and theologians mentioned in Chapter 5. But it doesn't mean we can't have fun trying.

17

INFINITY TO GO

Only two things are infinite, the universe and human stupidity, and I'm not sure about the former.

Attributed to Albert Einstein
(reputedly said at a press conference in the 1930s)

Space is almost infinite. As a matter of fact we think it is infinite.

Dan Quayle, *The Sunday Times*, 3 December 1989

THERE ARE, WE HAVE DISCOVERED, three ways of looking at mathematics. Each view is entirely sound, yet there are only partial overlaps between the three. The first is the world of pure mathematics, the great game, where it doesn't frankly matter whether or not a concept could ever have any application in the real world. Many of the great mathematicians of history have been most comfortable in this abstract universe, where the important thing is not to find a practical use for a theory but to discover effective and elegant methods with which theorems and proofs are slotted into place.

At the other extreme is real mathematics, the mathematics that would best suit Cantor's nemesis, Leopold Kronecker, where every mathematical concept has to be grounded in physical reality. Not only does this exclude the use of i, the square root of -1, but arguably it does not recognize the existence of negative numbers themselves. 'Can you show me -3 apples?' the real mathematician may enquire. Even zero as anything other than a placeholder in calculations is treated with some suspicion. Is there truly a real, physical 'nothing' anywhere? What does it look like or feel like? We can only ever see it as the absence of something.

It might seem such real mathematics ought to be the tool to be used by scientists and engineers. But in practice there's a third way. One that sits between real and pure mathematics – applied mathematics. This is the mathematics of the pragmatist. It is prepared to make use of the non-real elements of mathematics along the way, as long as you never end up with an unreal result. Now it's okay to have -3 apples, as long as you add them to 5 apples and end up with 2 real apples left. (But don't try to add -3 apples to 1 apple, please.) Similarly it's fine to stray into the square root of -1, provided that it is a means to an end, and you don't end up with a real, tangible object that has i as one of its dimensions.

Each of these viewpoints is taken by some mathematician or other. Here is the late philosopher Harvard Professor Willard Van Orman Quine on the subject:

> Pure mathematics, in my view, is firmly embedded as an integral part of our system of the world. Thus my view of pure mathematics is oriented strictly to application in empirical science. Parsons [Charles Parsons, Edgar Pierce Professor of Philosophy at Harvard University] has remarked, against this attitude, that pure mathematics extravagantly exceeds the needs of application. It does indeed, but I see these excesses as a simplistic matter of rounding out. We have a modest example of this process already in the irrational numbers: no measurement could be too accurate to be accommodated by a rational number, but we admit the extras to simplify out computations and generalizations. Higher set theory is more of the same. I recognize indenumerable infinites [i.e. infinites beyond $\aleph_0$, such as all the numbers in the continuum between 0 and 1] only because they are forced on me by the simplest known systematizations of more welcome matters.[108]

So where do infinity and the infinitesimally small fit? Clearly there is no reason why they should not be acceptable parts of pure mathematics. Provided that they can be shown to work in a consistent way with the set of axioms on which a particular mathematics is based, they are appropriate grist to the mill. But is there a real infinity that means that, unlike −3, this is a concept that can be encountered in the real world? Can we conceive of true infinitesimals? Or are all our uses of infinity and the infinitesimal confined to the pragmatic middle ground, to be ironed out before we return to reality?

The great eighteenth-century Scottish philosopher David Hume was convinced of the unreality of infinity. It might seem that philosophy should be a subject only undertaken after many years of experience of the human condition, but Hume was a natural prodigy. He had matriculated at the University of Edinburgh at the age of 12, and his great philosophical study, *A Treatise on Human Nature* was published in 1739 when he was only 28.

For Hume, the essential problem of infinity was our inability to actually envisage it:

> The capacity of the mind is not infinite, consequently no idea of extension or duration consists of an infinite number of parts or inferior ideas, but of a finite number, and these simple and indivisible...[109]

In other words, when we think of anything that has physical dimensions, or duration in time, we can only imagine it being divided up into a finite number of parts – there has to be some mental limit to our ability to subdivide. Hume does not even allow us to conceive of infinity by use of those handy three dots of the mathematician, unlike his English predecessor John Locke, who did the best of his work in the 1690s. Locke was equally sure we couldn't conceive of an infinite object, but was happy that we could cope with taking a finite chunk and repeating it over and over, with the idea that by never stopping we encompass infinity.

As he put it:

> [The infinity of space] is nothing but a supposed endless progression of the Mind, over what repeated ideas of Space it pleases; but to have actually in the Mind the *Idea* of a Space infinite, is to suppose the Mind already passed over and actually to have a view of all those repeated *Ideas* of Space...[110]

Locke, indeed, was prepared to admit (perhaps sarcastically) that others were capable of imagining things that he was not:

> But yet if after all this, there be men who persuade themselves that they have clear positive comprehensive ideas of infinity, it is fit that they enjoy this privilege: and I should be very glad (with some others that I know, who acknowledge they have none such) to be better informed by their communication.[111]

Hume did not harbour any such openness to persuasion. He based his viewpoint on a mixture of logical arguments and experiment. Though Hume's prime experiment was very subjective by modern standards, it is still worth examining. One of the joys of such an experiment is that it is one that anyone can carry out. For obvious reasons, it has become known as the ink-spot experiment.

> Put a spot of ink upon paper, fix your eye upon that spot, and retire to such a distance, that at last you lose sight of it; 'tis plain that the moment before it vanish'd the image or impression was perfectly indivisible. 'Tis not for want of rays of light striking our eyes, that the minutest parts of distant bodies convey not any sensible impression; but because they are remov'd beyond that distance, at which their impressions were reduc'd to a minimum, and were incapable of further diminution.[112]

Because the experiment is so simple, it is worth recreating Hume's effort from around 260 years ago. The spot certainly does disappear, and there is a fuzzy point at which something seems to come into being. In terms of the resolution of the eye, at least, it seems reasonable that the image of the ink spot has become indivisible – that to divide it any further would result in it disappearing (something you can prove, as Hume pointed out, by cutting the spot in half and having an assistant split and rejoin them).

Yet performing the experiment doesn't quite show what Hume wanted us to see. He claimed that what we see is a 'sensible, extensionless' item. It is sensible in that it is detectable by our senses, and Hume suggested it was extensionless – in effect had no measurable size – because to make it any smaller would make it disappear. But it is not clear that just because something is at the limits of visibility that it is truly extensionless. This is one of the lesser sticking points that emerge as we work through Hume's treatise. But next comes what is arguably the really shaky leap.

Hume believed that ideas were always in some sense reflections of reality. That they were 'faint images' or impressions of sensation or reflection. For Hume, every idea had to have as its basis some form of sense impression. It's easy to see why he thought this. Try to imagine a new colour, one that you've never seen, one that bears no resemblance to any other colour you have ever seen. We can describe the concept, but it's impossible to have a clear mental picture, an idea, of how such a colour looks. However, the danger is to extend, as Hume did, from the senses to reality. Although we can't envisage a new colour, the truth is there are plenty of colours we can't see – infrared and ultraviolet for beginners.

Hume extended this flawed view to cover infinity. Since, he argued, our ideas are limited to finite division, then so was reality. He believed that because we can't conceive of something divided into an infinite number of parts, then it really *is* impossible 'without any farther excuse or evasion'.[113] He gives us the specific example of a grain of sand. The idea of the grain, he says, is not capable of

> separation into twenty, much less into a thousand, ten thousand or an infinite number of different ideas.[114]

It's easy to counter this argument by imagining that we are viewing the sand grain through an electron microscope (or for that matter, seeing the ink spot through a telescope), but Hume hadn't missed this possibility. For him, the point is that, whatever your ability to resolve what you see into detail, in any particular situation there will be a limit beyond which you can't see, a division beyond which you can't envision the separate ideas to make it up.

A stronger argument against Hume's view is to attack not the scale of operation, but the leap that says what we can't divide mentally isn't capable of such division. We know that a human body, for example, can be divided up into its component atoms. No one could cope with a mental picture of every single atom in the body, yet it is possible to physically divide off one atom. Similarly none of us can form a true mental picture of every single person in the world – but it is very easy to divide off an individual person from the mass. The fact that we can't achieve such a division mentally does not make it physically impossible.

The way Hume links what we can perceive and reality may seem to be the fancying of a person writing in an age when very little was known about the brain or the fine detail of creation, but there are surprisingly strong echoes in a piece written by David Hilbert, one of the greatest twentieth-century mathematicians, who did not make any great breakthroughs in the mathematics of infinity, but helped clarify many of the details.

Hilbert was born at Königsberg in German (now Kaliningrad in Russia) in 1862 and by the turn of the century he was recognized as the pre-eminent European mathematician. Perhaps most famously he spoke at the Second International Congress of Mathematicians at Paris in 1900, defining the 23 greatest problems facing mathematicians. These included the famous Fermat's Last Theorem and number one on the list was Cantor's continuum hypothesis. But Hilbert's echoes of Hume were not in trying to deal with the continuum hypothesis or any other specific theory, but rather with infinity itself. He wrote in 1926:

> We have already seen that the infinite is nowhere to be found in reality, no matter what experiences, observations and knowledge

are appealed to. Can thought about things be so different from things? Can thinking processes be so unlike the actual process of things? In short, can thought be so far removed from reality? Rather is it not clear that, when we think that we have encountered the infinity in some real sense we have merely been seduced into thinking so by the fact that we often encounter extremely large and extremely small dimensions in reality.[115]

In hindsight there are two separable arguments in Hilbert's words. It may well be true that when we think we have encountered infinity we have in fact only encountered the extremely large or the extremely small, but his echo of Hume's assumption that thought and reality were interlinked has to be doubtful. After all, these days we are used to watching movies where apparently real computer-generated characters share the stage with real people. To suggest that thought processes are like the things thought about is like suggesting that these virtual people are in some way like real people in the way they are produced. The senses may tell us this, but in fact the model that drives them has a totally different basis, and it's hard not to take the same view of the apparent link between our ideas and reality. One is but a model of the other.

Professor Shaughan Lavine of the University of Arizona points out that Hilbert may not have been making a statement about thought as a modelling process, but rather saying that given that the infinite is found nowhere in reality, then it cannot intrude into thought either. As thought has a physical basis, Hilbert seems to argue, and as there can be no infinite in physical reality, then we can't cope with thinking about infinity either. Of course, this argument depends on having established that infinity cannot be part of reality (back to Hume).

Although Hume's arguments have themselves rather suffered with age, the principle of avoiding the use of infinity if at all possible is one that still has some value. Dale Jacquette, professor of philosophy at Pennsylvania State University, argues just as Cauchy did before him, that it is perfectly possible to operate all of calculus without ever bringing in the tricky infinitesimal. If infinity isn't necessary, he argues, why bother with it? This is a perfectly valid viewpoint when considering the needs of calculus, though it does not in any way demonstrate that infinity does not exist. Jacquette, clearly a fan of Hume's, points out that

> although Hume's critique [of infinity] is not universally condemned, the tide of opinion has overwhelmingly opposed [it] even when the objections brought against it are based on misinterpretations.[116]

According to Jacquette, we might not agree with all of Hume's arguments, but his position that we don't need infinity (whether or not it can actually be proved not to exist), and so shouldn't use such a complication and confusing concept in practical mathematics, still makes a lot of sense in the limited field of calculus. The idea is to replace the potential infinity ∞, with a new quantity – let's give it the symbol ₪. This is very specifically not infinity, not even a potential infinity, but is a sort of broken infinity with bits missing. It is an indefinite quantity, never actually determined, that is inexhaustible.

At first sight there's not much to choose between inexhaustible and infinite. Doesn't ₪ have to be infinite to be inexhaustible? No – that's the nice thing about it. We are never going to make use of more than a finite part of ₪, so we simply require it to be bigger than the biggest number we can possibly want. Inexhaustible, but finite. Professor Jacquette suggests we replace infinity with this inexhaustible figure – just take my ₪ character and plug it into all those places where ∞ is currently used.

Apart from the change of symbol, most of mathematics and science would continue unchanged, but we would have reduced the need to fight Bishop Berkeley's ghostly quantities. If you wanted a name for ₪, *apeiron* wouldn't be a bad label. This isn't quite the same as 'inexhaustible', but ₪ has, after all, something of the flavour of this ancient Greek word that was used for infinity, but also for indefinite. In fact Aristotle, when contemplating the nature of infinity in his *Physics*, came up with a surprisingly similar surmise to that of Professor Jacquette:

> Our account does not rob the mathematicians of their science, by disproving the actual existence of the infinite... In point of fact they do not need the infinite and do not use it. They postulate only that the finite straight line may be produced as far as they wish.[117]

This use of the inexhaustible follows on directly from Bernard Bolzano's criticism of the way that ∞ was used as a variable something that would be bigger than anything you could specify, but never quite reached the true, absolute infinity. In *Paradoxes of the Infinite* (see Chapter 10) Bolzano points out, as Professor Jacquette would later, that it is possible

> for a quantity merely capable of becoming larger than any one pre-assigned (finite) quantity, nevertheless to remain at all times merely finite.[118]

Bolzano intended this as a criticism of the way infinity was treated, but Jacquette sees it instead of a way of making use of practical applications like calculus without the need for weasel words about infinity.

By replacing ∞ with ℵ we do away with one of the most common requirements for infinity, but is there anything left that maps on to the real world? Can we confine infinity to that pure mathematical other world, where anything, however unreal, can be constructed, and forget about it elsewhere? Surprisingly, this seems to have been the view, at least at one point in time, even of Cantor himself. In a book written in 1883, he comments that only the finite numbers are real.

Perhaps the most natural place to look for infinity in the real world is in the extent of the universe. Even in Archimedes' day, when he tried to work out the number of grains of sand to fill the universe in the *Sand-reckoner* there must have been some sense of awe at the sheer scale of this quantity. Yet by modern standards, Archimedes' universe at around 16 million kilometres across was a tiny speck on the face of creation.

Through the centuries of the development of science, our picture of the size of the universe has been expanding. In the classical concept of the universe developed by the late Greek philosopher Ptolemy, where the Earth was at the centre of a series of spheres, the outermost being the one that carries the stars, this 'sphere of the fixed stars' (as opposed to the moving planets) began at 5 myriad myriad stades and 6,946 myriad stades and a third of a myriad stades. A myriad is 10,000 and each of the stades is around 180 metres long, amounting to around 100 million kilometres. Though it wasn't clear how thick this sphere was considered to be, it still is rather on the small side when you consider that the nearest star, Alpha Centauri, is actually around 4 light years, roughly 38 million million kilometres, away.

Copernicus not only transformed astronomy by putting the Sun at the centre of the Solar System, he expanded its scale, putting the sphere of the stars at around 9 billion kilometres. It wasn't until the nineteenth century that these figures, little more than guesses, were finally put aside when the technology had been developed sufficiently for the first reasonably accurate measurements to be made. The very first measurements, although all based on very near (in galactic terms) stars, made it clear that the stars varied considerably in distance, with one of the first stars measured, Vega, found to be more than six times as far away as Alpha Centauri – a difference in distance of a good 2×10^{14} kilometres – nothing trivial.

By the early twentieth century, the small Magellanic Cloud, now known to be a close satellite of our galaxy was found to be 30,000 light years away (compare this with the 26 light years of Vega), each light year being around 9,500 billion kilometres. In fact this too was an underestimate, as the Cloud is now thought to by closer to 169,000 light years distant, yet it shows how once again, the distances involved were on the move. By 1958, the extent of the universe was considered to be around 13 billion light years (around 12×10^{22} kilometres).

As more and more discoveries were made, it seemed quite

feasible that the universe went on forever. Why should it stop? After all, if the universe does have a boundary it presents real problems – what is outside the boundary? This was a problem that was already recognized in ancient times, when the Roman philosopher Lucretius had pointed out that either a dart thrown at the edge of the universe would carry on – in which case there wasn't an edge – or it would stop – in which case, what was outside the edge that was stopping it?

Roger Bacon, the medieval friar who summarized the scientific knowledge of his time in his *Opus majus*, believed that the universe was indeed finite and put forward a geometric proof of this.[119] He imagined drawing lines from the centre of the universe out to its edge (see Figure 17.1).

He starts with two lines (the dotted line and the lightly drawn complete line), starting from the same point at the centre of the universe and stretching to the edge of this immense sphere. Because his universe is spherical, these lines are each of the same length. He then adds a third line (the heavily drawn line in the diagram), parallel to the lightly drawn line and also heading off to the edge of the universe. Now if the universe is infinite in size,

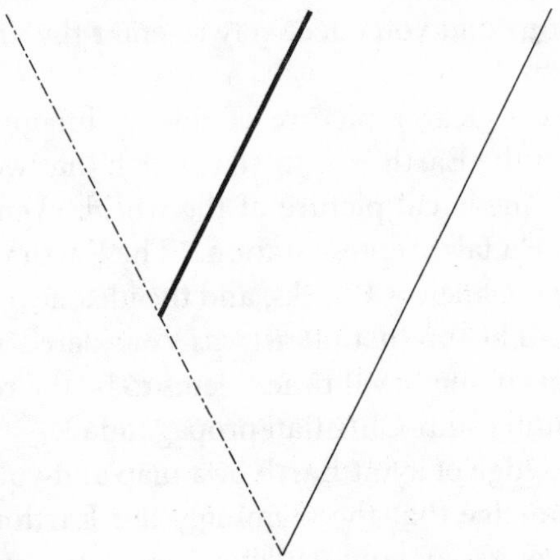

Figure 17.1   Bacon's proof of the finite universe.

he argued, the two parallel lines should be the same length (presumably because two parallel lines meet at infinity) and the dotted line the same length as the lightly drawn line.

But the heavy line is also the same length as the part of the dotted line up to where they meet – once more there are a pair of lines going from a point to infinity. So, argues Bacon, part of the dotted line equals the whole of the dotted line. This is an impossibility. So the edge of the universe isn't at an infinite distance at all (meaning that the parallel lines weren't the same length after all). Unfortunately, Bacon wasn't aware that adding a small quantity to infinity still produced infinity – but this was hardly surprising for a man writing in 1266.

Since Bacon's time, our ideas of the size of the universe has fluctuated between a true infinity (as Newton, for example, believed it to be), and the limits of measurement of the most distant cosmic objects. This latter view, which is the majority view at the moment, still prompts the question of what happens at the edge – if space is not infinite, what is beyond the boundary of the universe. The most popular suggestion is that there is no edge in the sense that we are familiar with. As you reach the limits of the universe, space is so folded in on itself (through some unspecified other dimension) that you effectively re-enter the universe at the far side.

You can get a clearer picture of this by imagining walking across a map of the Earth – or the flat Earth that we still tend to think of as the medieval picture of the world, even though this now seems to be a false representation. (The Earth was known to be a sphere by the ancient Greeks, and the idea never really went out of favour. The myth that it was considered flat even by educated people in medieval times seems to be the result of some nineteenth-century anti-Christian propaganda.)

Come to the edge of a flat Earth or a map and you fall off. But we know in practice that the seemingly flat Earth doesn't have an edge. You could carry on walking (assuming you have some mechanism for walking across a sea bed) forever without reaching the boundary. This is because the imaginary flat Earth is actually folded through a third dimension to make a sphere. Similarly, the picture of the universe that makes it realistically unbounded requires it to be folded through a fourth dimension.

Another possible physical manifestation of infinity comes when we consider the nature of time. We now believe that the universe began, probably from the Big Bang, around 15 billion years ago. But this does not mean that time started 15 billion years ago – we can't know if there was any time pre-Big Bang. Similarly we know of no mechanism for time to end – it seems reasonable to suppose it might go on forever.

Some cosmological theories put forward an end of the universe in a Big Crunch, an inverse of the Big Bang where all matter is brought together and annihilated, a point at which it is possible that time might end, just as it might have begun with the Big Bang. However, there are also cyclical theories that see the Big Crunch followed by another Big Bang, and so on for ever. Probably the best supported current theory sees the universe ending in slow heat death, running down without ever coming back together – and with no clear point at which time could be considered to finish, unless there is finally a state where the whole universe comes to rest.

It is hard to see how there could ever be a certain answer to the question of whether or not time is infinite, but there is one further approach to physical infinity that seems at first sight easier to cope with – the infinitely small. It might seem after taking in Cantor's elegant proof that there are the same, infinite number of points on a line and in a square or cube that this infinity, at least, has a physical reality. But Cantor's proof involves a hypothetical mathematical line, square or cube. The real thing need not be a continuum in the way Cantor's line is.

This is the argument Hume was trying to make with his inkspot experiment. Matter may not be infinitely divisible – there might be a limit, even if it isn't Hume's 'points' that could still be sensed. We know now, at the very least, that we can divide matter into atoms, that those atoms can be divided into the subatomic particles electrons, protons and neutrons, and possibly that these particles can themselves be divided into quarks. (I say possibly, because quarks have never been observed, and the quark theory may be simply a way of defining 'flavours' of particle, rather than representing true sub-components.)

Perhaps this division doesn't stop there, but can go on forever. Quantum mechanics has introduced a degree of uncertainty into our measurements, but doesn't actually prevent infinitely small 'particles' existing. One of the best known tenets of quantum mechanics, Heisenberg's uncertainty principle, is often misused to suggest that we can't ever know anything accurately. But in fact, the principle only says that your ability to measure position is linked to your ability to measure momentum (mass × velocity). The more accurately you know the position, the less certain you are about mass, speed and direction of movement. This doesn't mean that there is any concrete limit on our ability to measure one of these factors, nor does it imply that space can't be divided into indefinitely small quantities.

In practice, though, the nature of infinity means that we are never going to measure an infinitely small component of space. Because we have finite capabilities, there will always be some granularity, some limit to the practicality of measurement. But that does not prove that space (or time) is not infinitely divisible, simply that we can never practically achieve it.

It would seem the best bets are that the universe is not infinite, but that we can never be sure if time is infinite, nor can we necessarily be sure that it isn't possible to divide up space or time into an infinite number of infinitesimal components.

One field of modern mathematics that seems to open up the boundaries for infinitesimal space in the real world is fractals. This fascinating adjunct to chaos theory demonstrates quite painlessly that it is possible in principle to have something infinitely long drawn in normal, finite space. The simplest way to imagine this is by producing what is called a Koch curve, first described by

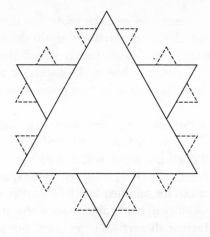

Figure 17.2   The beginnings of a Koch curve.

the Swedish mathematician Niels Fabian Helge von Koch in 1906.[120]

Von Koch, one of Gösta Mittag-Leffler's pupils at the University of Stockholm, began with a simple equilateral triangle. On each of the three equal-length sides he placed another equilateral triangle, exactly one third the size, and in the middle of the side (see Figure 17.2).

If you look at each of the three sides of the original triangle, it is now extended to be made up of four segments, each 1/3 the length of the original line – so the length of each of the three 'sides' is now 4/3 of what it was. Koch now went on to perform the same 1/3 side triangle addition to each of the smaller sub-sides. So now each of these is 4/3 longer. And this process is repeated forever. The sides get longer and longer, becoming infinitely long, but the shape, rather like a crinkly snowflake, does not grow beyond the bounds of a circle drawn around the original triangle.

Without going to the extremes of the Koch curve, this process demonstrates why it is in practice impossible to say exactly how far it is around a complex fractal shape like the outline of the British Isles. You can measure the distance on a map, but in the real world, the closer you get, the more fine detail you see (as Douglas Adams's character Slartibarfast, the designer of Norway in the *Hitch-hiker's Guide to the Galaxy* called them, 'the lovely crinkly edges'[121]). And that means the distance can be pretty well as long as you like, depending on how many of those crinkles you follow.

Fractals certainly allow us to envisage the possibility of infinity in a finite world, but we are back to the need to divide space up into infinitely small portions, which will never be practically possible even if it is theoretically. It is, however, in the quantum world of photons of light and the minute particles of matter that there is the most likely possibility of being able to see a true infinity at work, not in some abstruse theoretical experiment but as the potential basis for a totally new form of computer. One that makes use of infinity itself to achieve what had been assumed was impossible.

It has been known since 1936 that, however well we build them, computers should always have limits. This was proved by Alan Turing, best known as one of the code-crackers at the wartime Station X at Bletchley Park, and one of the fathers of modern computing. Turing showed that there were certain types of problem where the result could never be predicted. In an argument similar to Gödel's ideas on undecidability, Turing proved that that for some kinds of computer program that repeatedly attack a mathematical problem it would never be possible to predict whether or not the program would ever finish.[122]

This result is not just a fact about the workings of computers. Take, for example, another of the 23 great mathematical problems identified by David Hilbert in 1900, the Goldbach conjecture. This supposes the reasonable, but as yet unproved, idea that every even number greater than 2 can be made up by adding together two prime numbers. Primes are simply whole numbers that aren't divisible by anything else but themselves and 1. They feature regularly in mathematics, and Goldbach's unproved suggestion seems a reasonable one, but one that has not been possible to confirm by logic alone.

Although many mathematicians don't like it, computers can sometimes be used to *disprove* a conjecture like this. Through sheer brute force (rather than mathematical elegance), the program can work through number after number trying out the result. Such an approach can't prove that something is true, but if it ever finds a case where it's not true, the conjecture can be disposed of. In this case, if the computer hits on any even number that can't be made up by adding together two primes, Goldbach's conjecture would be proved false.

In principle it might have been possible, rather than to try out every number, to predict the outcome using some mathematical logic. But Turing's discovery showed that it wasn't *ever* possible to predict whether or not that computer would find an example that proved the Goldbach conjecture wrong. The only way to get round it would be to find a way to have a computer program that, instead of working through the problem sequentially, step by step, tried out every possible number – a whole infinite set of them – simultaneously. This clearly would never happen. Or so it seemed.

To find a way around Turing's 'halting problem', Cristian Calude and Boris Pavlov of the University of Auckland in New Zealand proposed in 2002 using one of the strange quirks of the quantum-mechanical world that we have already met in Chapter 3. When working with extremely small particles, normal physical expectations simply don't apply. It seems, for instance, that the strange behaviour of quantum mechanics makes it possible to communicate instantly across any distance. But the useful possibility here is that a particle can be in more than one state at the same time.

This is the crazy possibility that Schrödinger's cat demonstrated. In that imaginary experiment the cat's life depends on the state of a particle, which is a superposition of two different states until it is observed. Because the particle is thought to be in both possible states at once, so the cat is both alive and dead simultaneously. If such superposed states do exist – and they are

accepted by most physicists these days, even though the practicalities of the Schrödinger's cat experiment are not – then a theoretical mechanism has been devised to make use of such
simultaneous states of being to perform a process called quantum
computing, using the different states as active bits to perform
massive simultaneous computations. Of itself, this doesn't get
round Turing's problem, but here at last is where infinity may –
just may – come into the real world.

Calude and Pavlov point out[123] that an atom like hydrogen has
an infinite number of possible states, each with a different level of
energy. It's as if the atom is a battery, and each state has a little
more energy crammed into it. Yet this is despite the atom (again
like a battery) having a finite limit to its energy content. All the
levels are fitted into a finite 'space' by a process just like Zeno's
paradox – each state is of higher energy than the last, but the
extra amount of energy involved gets smaller and smaller. The
proposal from the New Zealand scientists is to use a superposition
of all these infinite number of states as if they were an infinite
number of computer bits, programming it to attack every single
possible solution at once and find out (for instance) if every even
number can be made up from two primes.

In practice this would not provide an absolute solution to the
problem. Working at the quantum level, nothing is straightforward. It isn't possible to directly measure the outcome as this
would change the result. With the indirect measurements that are
possible, Calude and Pavlov's method could only produce a probability of whether or not Goldbach's conjecture (or any other
problem requiring such an infinite series of checks) is true, with
the ability to get a better and better probability depending on
how long the process is run. But that's one step better than the
current position where there is no probability attached to the
outcome. Here, at least in principle, is a real result dependent on
a 'real' infinity.

Have we proved, then, that infinity can ever escape the theoretical constructs of mathematics into the real world? No one can

say for certain until such quantum computers are built and tested. Most mathematicians would probably argue that it doesn't really matter − it is much more important to have an interesting challenge than it is to worry about applicability. There is no mathematical certainty about the existence of the infinite in the universe, but equally there seems to be no good reason for believing that the existence of the physically infinite is an impossibility. The physical realities of infinity must remain uncertain, yet for the pragmatist, infinity is a valuable part of the mathematical toolkit. It's useful. It works.

Usefulness isn't everything, though. Talk to mathematicians or scientists and you will hear not only of practical solutions, but of *elegant* solutions. Ideas and theories that have a beauty all of their own. When Richard Feynman was introducing the concepts of quantum theory in his book *QED*, he remarked:

> So I hope you can accept Nature as She is − absurd.
>
> I'm going to have fun telling you about this absurdity, because I find it delightful . . . and I hope you will be as delighted as I am when we're through.[124]

Feynman is not telling us how useful this theory is (though it indubitably is useful), but how *delightful*. It doesn't just make us better informed or educated, it should be filling us with delight. Again, when the thirteenth-century friar Roger Bacon was writing about his favourite aspect of science, light, it was clear that he was dealing with more than simple practicality:

> It is possible that some other science may be more useful, but no other science has so much sweetness and beauty of utility. Therefore it is the flower of the whole of philosophy . . .[125]

Feynman and Bacon each expressed a sense of enjoyment, of adventure in discovering the remarkable secrets of their own fields of discovery. In the final, short chapter we will use a pair of paradoxes to explore the way that the subject of infinity fascinates us all.

18

ENDLESS FASCINATION

So, naturalists observe, a flea
Hath smaller fleas that on him prey;
And these have smaller fleas to bite 'em;
And so proceed ad infinitum.

Jonathan Swift, *On Poetry*

OFTEN, THE BOUNDARIES BETWEEN reality and infinity throw up paradoxical results, and it is these mind-teasing outcomes that give the study of the infinite a playfulness and fascination that lives on, however much we explore the philosophical, mathematical and theological implications of the subject.

Galileo might have been using the concepts of infinity to explore the realities of the physical world, but it is obvious when reading his book just how much he was also enjoying the paradoxes thrown up by his linked bigger and smaller circles that could traverse the same distance in a turn, or the confusion arising from the fact that there is the same infinite quantity of whole numbers and squares. A similar playfulness seems to have influenced David Hilbert, the mathematician behind the 23 unsolved problems at the 1900 Congress, when he dreamed up a very strange hotel.

Hilbert's hotel is not one that you can expect to encounter in the average seaside resort. Hilbert imagined that he had a building with an infinite number of rooms. This expansive concept was to be echoed in a segment of Douglas Adams's *Hitch-hiker's Guide to the Galaxy*, where a restaurant is sited at the end (in time) of the universe. The fictional restaurant exists in a time bubble that is repeatedly allowed to drift up to the final destruction of the universe. Adams imagines a restaurant with an infinite capacity by manipulating time, effectively setting up an infinite number of parallel restaurants.

Hilbert did not bother to construct fictional logic for his hotel – it was more an exercise in exploring the paradoxical impact of infinity. And the hotel displays some particular entertaining peculiarities. It's not just an attractive proposition for the hotelier because an infinite number of rooms suggests the possibility of infinite revenue. It also means that he never need worry about finding spare capacity, no matter how full the hotel is or how many people turn up.

Imagine that Hilbert's hotel is entirely full. Every single room is occupied, down to the pokey little spaces in the attic. A late arrival scrambles into the foyer. Can they put him up? No problem (though it will take some cooperation from the other guests). The management moves each person into the next room up. The occupant of number 1 goes into number 2, the resident of number 2 into room 3, and so on. Room 1 is vacated, but no one leaves.

Hilbert even imagined an infinitely large coach load of visitors arriving unexpectedly. It's bad enough to have a conventional coach, but this one (thanks to some strange manipulation of reality) manages to cram in an infinite number of people looking for bed and board. Even now, though, the owner of Hilbert's hotel doesn't have a problem. The enthusiastic proprietor, rubbing his hands and imagining the impact on his bank balance, simply moves everyone into the room with a number twice as large as their current one. So, 1 goes to 2, 2 to 4, 3 to 6, and so on. Suddenly all the odd numbered rooms – an infinity of them – are available for the coach occupants.

Hilbert is playing on the same point that Galileo was aware of when he matched up numbers and their squares. There is no difference between aleph-null infinities,

$$\aleph_0 + 1 = \aleph_0 \quad \text{and} \quad \aleph_0 \times 2 = \aleph_0,$$

so his hotel is always ready to take on extra guests. (This also means that the hotelier is wasting his time, as he can't increase his income once he has reached $\aleph_0$ guests.) But Hilbert demonstrates neatly how applying the mathematics of infinity to the real world makes for a remarkable picture.

It doesn't matter that this is an example of the infinite that could not possibly be achieved. After all, there are simple physical restrictions that force this to be the case. We live in a universe with a finite number of atoms – estimated loosely around 10^{80} – and that seems to have been in existence for a finite time – around 15 billion years. There isn't space for that infinite number of rooms, nor is there time to shift everyone into place. Yet the image of Hilbert's hotel remains a fascinating one.

This infinitely elastic hotel illustrates an infinity-based paradox that is very familiar to us at this stage of the book, but even with a good grasp of the workings of the infinite, some paradoxes remain surprising. There are few better examples of the infinite's mix of simplicity and mind-boggling impossibility than the mathematical construction that is dubbed Gabriel's horn.

This is the result of turning into three dimensions the outcome of a simple mathematical function $f(x) = 1/x$ for every x greater than 1. Remember, the function simply turns one number into another. If $x = 2$, the function produces $\frac{1}{2}$. If $x = 3$, then $f(x)$ is $\frac{1}{3}$, and so on. This flat graph is turned into a three-dimensional shape by spinning it around the axis, like a lathe turning an object that is going to be carved. The object that results disappears off to a point, resembling a wizard's hat with an infinitely tall peak that gets narrower and narrower (see Figure 18.1).

At first sight this is like any other shape, but it has a very special

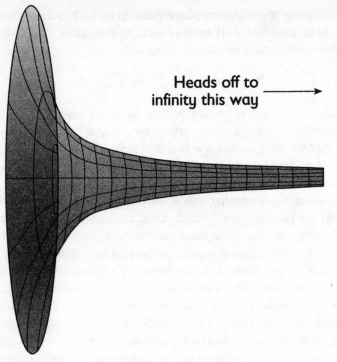

Heads off to
infinity this way $\longrightarrow$

Figure 18.1 Gabriel's horn.

property. Just as that familiar sequence we started with, $1, \frac{1}{2}, \frac{1}{4}, \frac{1}{8}$, $\frac{1}{16}, \frac{1}{32}, \ldots$ can go on for an infinite number of elements and still add up to a finite value, so the volume of Gabriel's horn is a finite amount. The horn might go on for ever, but just like the sequence, the cross-section gets smaller and smaller in a way that tends towards a limit. In the case of Gabriel's horn that limit is π rather than 2, but the result is similar. The volume of Gabriel's horn is always π.

You may be wondering how anything solid could have the volume π. Pi lots of what, exactly? Could it be π cubic metres or π cubic feet or π cubic centimetres? The answer is, yes – it could be any of them. Remember that this horn started off as a function of $1/x$ for every x that was greater than 1. So it's π lots of whatever we began with 1 of. If we start at 1 metre, the volume will be π cubic metres, and so on.

Of itself, having a finite volume despite being infinitely deep isn't too much of a surprise, but the truly amazing thing about Gabriel's horn comes when you work out not its volume, but its surface area. By using integration, part of the calculus originated by Newton and Leibniz, it is possible to show that the area of the surface of this horn is infinite. The mathematics is a little fiddly, but the result is not open to question. Whereas the volume can be likened to a series like $1 + \frac{1}{2} + \frac{1}{4} + \frac{1}{8} + \frac{1}{16} + \frac{1}{32} + \cdots$ that heads for a particular value, the size of the surface has no limit.

So here is a shape, a very simple shape, that has a finite volume but an infinite surface area. Now imagine filling the horn with paint. It's easy enough to do. We just turn it point down and pour the right amount of paint into it, π units, filling the horn right to the brim. Yet if we tried to take a paint brush and paint the whole surface of the horn we would never finish – there's an infinite area to cover. Think about it. You can fill the horn's entire volume, which surely would coat the entire inner surface, with a finite quantity, but you can't cover that inner surface with paint.

This paradox is often explained by saying that Gabriel's horn has a finite volume and an infinite area, and since area and volume aren't the same thing this is not a problem – but frankly this isn't an explanation at all. I have yet to see a good explanation of what *is* happening. But in fact it is possible to work out a physical picture of the paradox. To understand the paradox of Gabriel's horn we have to think of a real situation. There is a difference between the volume of a container and the amount of liquid (or anything else) you can put in it.

Imagine for a moment an ordinary coffee jar. I could fill it with paint, and the paint would fill up the entire interior. But if I put something in it that was large compared with the width of the jar, say hen's eggs, I wouldn't be able to fill it up. There would be gaps

I could never get another egg into. In the case of Gabriel's horn, as the horn gets narrower it too will become similar in width to the molecules of paint we are trying to pour in.

Imagine pouring a large can of paint into Gabriel's horn. You could fill it up to the top certainly, but you couldn't get paint right down to the infinitely distant point. After a while, the horn gets so narrow that the paint won't go any further. This is true whether you're pouring it in or painting the inside surface. Because as soon as the horn is narrower than the size of a paint molecule you can't get any further down it. This will happen in a finite distance. To paint all the surfaces up to that point will only take a finite amount of paint – less in fact than the volume of the horn.

In principle, if you tried to paint the outside of the horn you might need an infinite amount of paint – but in practice again you would reach a point where the horn was too narrow for paint molecules to stick around the outside and coat it. Once more you could only paint a finite portion of the surface, even though the area is in practice infinite. The horn truly does have an infinite surface area and a finite volume. It will hold a finite amount of paint (somewhat less than π units). But we can only ever get a finite amount of paint on that infinite surface.

Hilbert's hotel and Gabriel's horn illustrate for me why infinity is and always will be such a wonderful subject. Such paradoxes are simple enough for anyone to grasp, yet each can make you shake your head in amazement. There's something of reality about them, but also something unearthly – attributing the horn to an archangel seems just the right thing to do. Infinity is strange, powerful and outside our ability to fully comprehend, but it's also fun.

As long as we have a sense of wonder, a fascination with the nature of reality and the paradoxes of number, infinity will continue to stimulate philosophers and six-year-olds alike. Next time you consider the grandeur of the universe, the fine detail of matter, or simply the passage of time, remember to include the impact of the infinite. It is pretty well impossible to think about philosophy, theology or mathematics without this enduringly delightful concept adding to the experience.

When it comes to infinity, the possibilities are, perhaps inevitably, endless.

REFERENCES

1. Douglas Adams, *The Hitch-hiker's Guide to the Galaxy – The Original Radio Scripts* (Pan, 1985).

2. Richard Feynman, *QED: The Strange Theory of Light and Matter* (Penguin, 1990).

3. *Brewer's Dictionary of Phrase and Fable*, Millennium Edition (Cassell, 1999).

4. *Ibid.*

5. David Fowler, *The Mathematics of Plato's Academy* (OUP, 1999).

6. Reviel Netz, *The Shaping of Deduction in Greek Mathematics* (CUP, 1999).

7. Aristotle, *Physics*, trans. R. P. Hardie and R. K. Gaye (OUP, 1962).

8. *Ibid.*

9. *Ibid.*

10. *Ibid.*

11. *Ibid.*

12. Roger Bacon, *Opus majus*, trans. Robert Belle Burke (Kessinger, 1998).

13. Robert Kaplan, *The Nothing That Is* (OUP, 1999).

14. Plutarch, 'Marcellus', quoted in: *Introduction of The Works of Archimedes*, trans. T. L. Heath (Dover, 2002; reproduction of CUP, 1897).

15. T. L. Heath, *Introduction to The Works of Archimedes* (Dover 2002; reproduction of CUP, 1897).

16. *Ibid.*

17. *Ibid.*

18. *Ibid.*

19. *Ibid.*

20. *Ibid.*

21. *Ibid.*

22. *Ibid.*

23. *Ibid.*

24. *Ibid.*

25. *Ibid.*

26. John Donne, *The Sermons of John Donne* (University of California Press, 1953–62).

27. *Ibid.*

28. *The Oxford Book of English Mystical Verse* (OUP, 1917).

29. Macedonius, *Letter to Augustine*, AD 413/4, trans. James J. O'Donnell.

30. St Augustine, *The City of God*, trans. Marcus Dods, Book XI.30 (The Modern Library, 2000).

31. *Ibid.*, Book XI.31.

32. *Ibid.*, Book XII.11.

33. *Ibid.*, Book XII.18.

34. *Ibid.*, Book XII.18.

35. *Ibid.*, Book XII.18.

36. St Thomas Aquinas, *Summa theologiae* (Christian Classics, 1989).

37. *Bhagavad Gita*, trans. Ramanand Prasad, Chapter 11 (American Gita Society, 1988).

38. Amir Aczel, *The Mystery of the Aleph* (Pocket Books, 2000).

39. Galileo Galilei, *Dialogues Concerning Two New Sciences*, trans. Henry Crew and Alfonso de Salvio, Day One (Dover, 1954).

40. Shaughan Lavine, *Understanding the Infinite* (Harvard University Press, 1994).

41. Brian Clegg, *First Scientist: A Life of Roger Bacon* (Constable & Robinson, 2003).

42. Roger Bacon, *Roger Bacon's Letter Concerning the Marvellous Power of Art and of Nature and Concerning the Nullity of Magic*, trans. Tenney L. Davis (Kessinger, 1940).

43. Aristotle, *Noctium Atticarum de collatione sapientum*, quoted by Bacon in Ref. 11 above.

44. Galileo Galilei, *Dialogues Concerning Two New Sciences*, trans. Henry Crew and Alfonso de Salvio, Day One (Dover, 1954).

45. J. N. Crossley and A. S. Henry, *Thus spake al-Khwarizmi*: a translation of the text of Cambridge University Library ms. Ii.vi.5, *Historia Mathematica* **17**(2) (1990), 103–131.

46. *New Shorter Oxford Dictionary* (OUP, 1997).

47. J. Gies and F. Gies, *Leonard of Pisa and the New Mathematics of the Middle Ages* (New Classics Library, 1983).

48. Michael Stifel, *Arithmetica integra*.

49. *Ibid.*

50. Quoted in: Martin Gardiner, *Mathematical Puzzles and Diversions* (Penguin, 1965).

51. Roger Bacon, *Opus majus*, trans. Robert Belle Burke (Kessinger, 1998).

52. J. F. Scott, *The Mathematical Work of John Wallis* (Taylor & Francis, 1938).

53. John Wallis, *De sectionibus conicus, nova methodo expositis. Ibid.*

54. Carl Friedrich Gauss, Letter to Danish astronomer Heinrich Christian Shumacher, 12 July 1831.

55. Quoted by Antonio Favaro in the introduction to: Galileo Galilei, *Dialogues Concerning Two New Sciences*, trans. Henry Crew and Alfonso de Salvio, Day One (Dover, 1954).

56. Galileo Galilei, *Op. cit.*, Day One.

57. *Ibid.*

58. C. A. Bretschneider, *Die Geometrie und die Geometer vor Euklides*, trans. T. L. Heath (Leipzig, 1870).

59. Eli Maor, *e: The Story of a Number* (Princeton University Press, 1994).

60. Gottfried Wilhelm Leibniz, *De quadratura*, ed. Eberhard Knobloch (Abh. Akad. Wiss., 1993).

61. J. F. Scott, *The Mathematical Work of John Wallis* (Chelsea, 1981).

62. Isaac Newton, *The Correspondence of Isaac Newton*, Vol. 3, ed. H. Turnbull (CUP, 1961).

63. Isaac Newton, *Op. cit.*, Vol. 1, ed. H. Turnbull (CUP, 1959).

64. Isaac Newton, *Op. cit.*, Vol. 5, ed. Rupert Hall and Laura Tilling (CUP, 1975).

65. *Ibid.*

66. Joseph Stock, *Life of Bishop Berkeley* (London, 1776).

67. *Ibid.*

68. George Berkeley, *The Analyst: A Discourse Addressed to an Infidel Mathematician*, Sect. 1. In: *Works* (Nelson, 1949–58).

69. *Ibid.*, Sect. 2.

70. *Ibid.*, Sect. 3.

71. *Ibid.*, Sect. 5.

72. *Ibid.*, Sect. 5.

73. *Ibid.*, Sect. 7.

74. *Ibid.*, Sect. 13.

75. *Ibid.*, Sect. 35.

76. Ian Stewart, *From Here to Infinity* (OUP, 1996).

77. Bernard Bolzano, *Paradoxes of the Infinite*, trans. D. A. Steele (Routledge and Kegan Paul, 1950).

78. *Ibid.*

79. *Ibid.*

80. Quoted in: Eli Maor, *e: The Story of a Number* (Princeton University Press, 1994).

81. Bernard Bolzano, *Op. cit.*

82. Quoted in the introduction to: Bernard Bolzano, *Op. cit.*

83. Seymour Lipshutz, *Set Theory and Related Topics* (MGraw-Hill, 1998).

84. Georg Cantor, *Contribution the Founding of the Theory of Transfinite Numbers* (Dover, 1955).

85. *Dictionary of Scientific Biography* (New York, 1970–90).

86. William Dunham, *The Mathematical Universe* (Wiley, 1997).

87. Quoted in: Amir Aczel, *The Mystery of the Aleph* (Pocket Books, 2000).

88. Anon., *The queen's role*, www.royal.gov.uk.

89. Nobel Prize Committee, *Nobel e-museum*, www.nobel.se.

90. Quoted in: Amir Aczel, *The Mystery of the Aleph* (Pocket Books, 2000).

91. John Lasseter, Andrew Stanton, Peter Docter, Joe Ranft, Joss Whedon, Joel Cohen, Alec Sokolow, *Toy Story* (Disney/Pixar, 1995).

92. Galileo Galilei, *Dialogues Concerning Two New Sciences*, trans. Henry Crew and Alfonso de Salvio, Day One (Dover, 1954).

93. M. Fierz, *Girolamo Cardano, 1501–1576: Physician, Natural Philosopher, Mathematician, Astrologer, and Interpreter of Dreams* (Birkhäuser, 1983).

94. Amir Aczel, *Op. cit.*

95. Revelation 1: 8, *New English Bible* (OUP/CUP, 1972).

96. Lewis Caroll, *The Annotated Alice*, ed. Martin Gardner (Penguin, 1970).

97. Michael Stifel, *Arithmetica Integra*.

98. Georg Cantor, *Contributions to the Founding of the Theory of Transfinite Numbers* (Dover, 1955).

99. *Nature*, 416, 600 (11 April 2002).

100. *Ibid.*

101. *New Scientist*, vol. 174, issue 2338 (13 April 2002).

102. *Ibid.*

103. Georg Cantor, letter quoted in: Joseph Dauben, *Georg Cantor: His Mathematics and Philosophy of the Infinite* (Princeton University Press, 1990).

104. Georg Cantor, *Op. cit.*

105. Quoted in: Howard Eves, *Mathematical Circles Adieu* (PWS Publishers, 1977).

106. Gottfried Wilhelm Leibniz, *Mémoire de M. G.W. Leibniz touchant son senti-ment sur le calcul différentiel* (Paris, 1701). Quoted in: Eberhard Knobloch, *Arch. Hist. Exact Sci.*, 54 (1999).

107. K. D. Stroyan and W. A. J. Luxemburg, *Introduction the Theory of Infinites-imals* (Academic Press, 1976).

108. Willard Quine, *The Philosophy of W.V. Quine* (Open Court, 1986).

109. David Hume, *A Treatise of Human Nature* (Clarendon Press, 1978).

110. John Locke, *An Essay Concerning Human Understanding* (Clarendon Press, 1975).

111. *Ibid.*

112. David Hume, *Op. cit.*

113. *Ibid.*

114. *Ibid.*

115. David Hilbert, *Über das Undendlich*, trans. Stefan Bauer-Mengelberg, *Mathematische Annalen* 1926.

116. Dale Jacquette, *David Hume's Critique of Infinity* (Brill, 2000).

117. Aristotle, *Physics*, trans. R. P. Hardie and R. K. Gaye (OUP, 1962).

118. Bernard Bolzano, *Paradoxes of the Infinite*, trans. D. A. Steele (Routledge and Kegan Paul, 1950).

119. Roger Bacon, *Opus majus*, trans. Robert Belle Burke (Kessinger, 1998).

120. Helge von Koch, Une méthode géométrique élémentaire pour l'étude de certaines questions de la théorie des courbes plane. *Acta Mathematica*, vol. 30 (1906), 145–174.

121. Douglas Adams, *The Hitch-hiker's Guide to the Galaxy*, the original radio scripts (Pan, 1985).

122. Ian Stewart, *From Here to Infinity* (OUP, 1996).

123. *MIT Quantum Information Processing Journal*, vol. 1.

124. Richard Feynman, *QED: The Strange Theory of Light and Matter* (Penguin, 1990)

125. Roger Bacon, *Op. cit.*

INDEX